Materials Recycling

Materials Recycling

Edited by
Abel Hall

■ Larsen & Keller
www.larsen-keller.com

Materials Recycling
Edited by Abel Hall
ISBN: 978-1-63549-176-0 (Hardback)

© 2017 Larsen & Keller

▤ Larsen & Keller

Published by Larsen and Keller Education,
5 Penn Plaza,
19th Floor,
New York, NY 10001, USA

Cataloging-in-Publication Data

Materials recycling / edited by Abel Hall.
   p. cm.
Includes bibliographical references and index.
ISBN 978-1-63549-176-0
1. Recycling (Waste, etc.). 2. Waste products--Recycling. 3. Materials handling. I. Hall, Abel.
TD794.5 .M38 2017
628.445 8--dc23

The publisher's policy is to use permanent paper from mills that operate a sustainable forestry policy. Furthermore, the publisher ensures that the text paper and cover boards used have met acceptable environmental accreditation standards.

Printed and bound in the United States of America.

For more information regarding Larsen and Keller Education and its products, please visit the publisher's website www.larsen-keller.com

Table of Contents

Preface   **VII**

Chapter 1 **Introduction to Recycling**   **1**

Chapter 2 **Various Materials Recycling Methods**   **27**
    i.   Computer Recycling   27
    ii. Retrocomputing   75
    iii. Greening   93

Chapter 3 **Various Materials Recycling Strategies**   **134**
    i.   Kerbside Collection   134
    ii. Single-stream Recycling   141
    iii. Materials Recovery Facility   143
    iv. Source Reduction   145

Chapter 4 **Materials Recycling Codes: An Integrated Study**   **147**
    i.   Recycling Codes   147
    ii. Resin Identification Code   147
    iii. Waste Hierarchy   150

Chapter 5 **Sustainable Recycling of Energy**   **155**
    i.   Battery Recycling   155
    ii. Rechargeable Battery   161
    iii. Energy Storage   173
    iv. Renewable Energy   191

Chapter 6 **An Overview of Nutrient Cycle**   **221**
    i.   Nutrient Cycle   221

Permissions

Index

Preface

In today's world, population is on a rapid rise and the level of pollution is rising every day. Thus, materials recycling is the concept which is very important in the present scenario. It refers to the process of converting the old waste material into some new and useful product. This process helps in reducing different types of pollution like greenhouse gas emissions, water pollution, etc. It also reduces the consumption of fresh raw material, energy wastage, and also helps reduce waste. This book is compiled in such a manner, that it will provide in-depth knowledge about the theory and methodology of materials and recycling. The topics covered in this extensive book deal with the core subjects of materials recycling. This book unfolds the innovative aspects of materials recycling which will be crucial for the holistic understanding of the subject. This textbook will serve as a reference to a broad spectrum of readers.

A short introduction to every chapter is written below to provide an overview of the content of the book:

Chapter 1 - Conservation efforts across the globe use the motto of "reduce, reuse, and recycle" to stress the necessity to save the environment. Recycling refers to process of converting waste material into reusable objects to reduce waste of useful material, prevent consumption of fresh raw materials etc. This chapter provides a comprehensive overview of the topic of recycling and its importance in current times; **Chapter 2** - This chapter divides recycling methods in three broad categories-computer recycling, retrocomputing and greening. Computer recycling techniques detailed in the chapter include data shredder, data erasure, degaussing, data remanence, electronic waste and file deletion. Retrocomputing techniques like home computer remake, minimalism (computing) and abandonware have been discussed. Greening includes methods like community greens, green alliance, community gardening, village green, park and common land. All these methods and their contribution to the recycling efforts have been comprehensively studied in the chapter; **Chapter 3** - Some strategies that aid in the recycling of materials are curbside collection, single-stream recycling, materials recovery facility and source reduction. All these strategies aim at segregating garbage into categories of biodegradable and non-biodegradable, increasing awareness about waste generation and reducing volumes of garbage. This chapter provides details about each of these strategies and their importance; **Chapter 4** - To identify the material from which an item is made and to facilitate faster and easier recycling, recycling codes are added to items. This chapter provides a comprehensive account of all the recycling codes in use, resin identification codes on plastic products and waste hierarchy. This content also helps the reader understand how useful this nomenclature and cataloguing is when recycling products; **Chapter 5** - Recycling is very important in the case of items like batteries as they cannot be degraded and pose an environmental hazard due to the toxic constituents present. This chapter focuses on the topic of battery recycling, rechargeable batteries and energy storage. It provides an in-depth account of the types of batteries and their ease of recycling. There is a section dedicated to renewable energy as well. The topics discussed in the chapter are of great importance to broaden the existing knowledge on recycling; **Chapter 6 -** This chapter explores the topic of nutrient recycling which is a natural method of exchange of organic and inorganic matter back into production of living matter. Important in this is the phenomenon of nutrition transfer among different organisms. The content provides a comprehensive outline of the topics of complete and closed loop, ecosystem engineers and ecological recycling.

I extend my sincere thanks to the publisher for considering me worthy of this task. Finally, I thank my family for being a source of support and help.

Editor

Introduction to Recycling

Conservation efforts across the globe use the motto of "reduce, reuse, and recycle" to stress the necessity to save the environment. Recycling refers to process of converting waste material into reusable objects to reduce waste of useful material, prevent consumption of fresh raw materials etc. This chapter provides a comprehensive overview of the topic of recycling and its importance in current times.

Recycling is the process of converting waste materials into reusable objects to prevent waste of potentially useful materials, reduce the consumption of fresh raw materials, energy usage, air pollution (from incineration) and water pollution (from landfilling) by decreasing the need for "conventional" waste disposal and lowering greenhouse gas emissions compared to plastic production. Recycling is a key component of modern waste reduction and is the third component of the "Reduce, Reuse and Recycle" waste hierarchy.

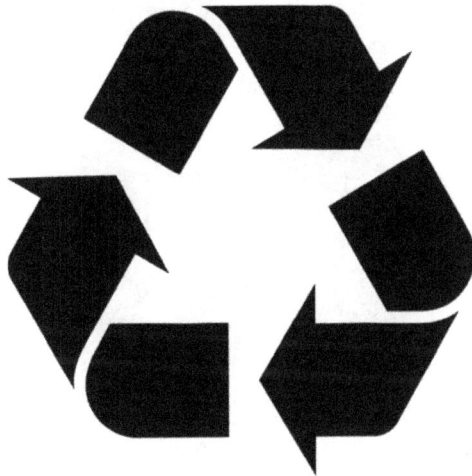

The three chasing arrows of the international recycling logo. It is sometimes accompanied by the text "reduce, reuse and recycle".

There are some ISO standards related to recycling such as ISO 15270:2008 for plastics waste and ISO 14001:2004 for environmental management control of recycling practice.

Recyclable materials include many kinds of glass, paper, metal, plastic, tires, textiles and electronics. The composting or other reuse of biodegradable waste—such as food or garden waste—is also considered recycling. Materials to be recycled are either brought to a collection centre or picked up from the curbside, then sorted, cleaned and reprocessed into new materials destined for manufacturing.

In the strictest sense, recycling of a material would produce a fresh supply of the same material—for example, used office paper would be converted into new office paper, or used polystyrene foam into new polystyrene. However, this is often difficult or too expensive (compared with producing the same product from raw materials or other sources), so "recycling" of many products or materials involves their *reuse* in producing different materials (for example, paperboard) instead. Another form of recycling is the salvage of certain materials from complex products, either due to their intrinsic value (such as lead from car batteries, or gold from circuit boards), or due to their hazardous nature (e.g., removal and reuse of mercury from thermometers and thermostats).

History

Origins

Recycling has been a common practice for most of human history, with recorded advocates as far back as Plato in 400 BC. During periods when resources were scarce, archaeological studies of ancient waste dumps show less household waste (such as ash, broken tools and pottery)—implying more waste was being recycled in the absence of new material.

An American poster from World War II

In pre-industrial times, there is evidence of scrap bronze and other metals being collected in Europe and melted down for perpetual reuse. Paper recycling was first recorded in 1031, when Japanese shops sold repulped paper. In Britain dust and ash from wood and coal fires was collected by "dustmen" and downcycled as a base material used in brick making. The main driver for these types of recycling was the economic advantage of obtaining recycled feedstock instead of acquiring virgin material, as well as a lack of public waste removal in ever more densely populated areas. In 1813, Benjamin Law developed the process of turning rags into "shoddy" and "mungo" wool in Batley, Yorkshire. This material combined recycled fibers with virgin wool. The West Yorkshire shoddy industry in towns such as Batley and Dewsbury, lasted from the early 19th century to at least 1914.

Industrialization spurred demand for affordable materials; aside from rags, ferrous scrap metals were coveted as they were cheaper to acquire than virgin ore. Railroads both purchased and sold scrap metal in the 19th century, and the growing steel and automobile industries purchased scrap in the early 20th century. Many secondary goods were collected, processed and sold by peddlers who scoured dumps and city streets for discarded machinery, pots, pans and other sources of metal. By World War I, thousands of such peddlers roamed the streets of American cities, taking advantage of market forces to recycle post-consumer materials back into industrial production.

Beverage bottles were recycled with a refundable deposit at some drink manufacturers in Great Britain and Ireland around 1800, notably Schweppes. An official recycling system with refundable deposits was established in Sweden for bottles in 1884 and aluminum beverage cans in 1982; the law led to a recycling rate for beverage containers of 84–99 percent depending on type, and a glass bottle can be refilled over 20 times on average.

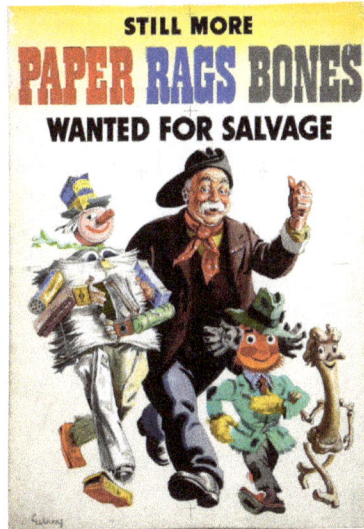

British poster from World War II

Wartime

New chemical industries created in the late 19th century both invented new materials (e.g. Bakelite (1907) and promised to transform valueless into valuable materials. Proverbially, you could not make a silk purse of a sow's ear—until the US firm Arhur D. Little published in 1921 "On the Making of Silk Purses from Sows' Ears", its research proving that when "chemistry puts on overalls and gets down to business . . .new values appear. New and better paths are opened to reach the goals desired."

Recycling was a highlight throughout World War II. During the war, financial constraints and significant material shortages due to war efforts made it necessary for countries to reuse goods and recycle materials. These resource shortages caused by the world wars, and other such world-changing occurrences, greatly encouraged recycling.

The struggles of war claimed much of the material resources available, leaving little for the civilian population. It became necessary for most homes to recycle their waste, as recycling offered an extra source of materials allowing people to make the most of what was available to them. Recycling household materials meant more resources for war efforts and a better chance of victory. Massive government promotion campaigns were carried out in the home front during World War II in every country involved in the war, urging citizens to donate metals and conserve fiber, as a matter of patriotism.

Post-war

A considerable investment in recycling occurred in the 1970s, due to rising energy costs. Recycling aluminum uses only 5% of the energy required by virgin production; glass, paper and metals have less dramatic but very significant energy savings when recycled feedstock is used.

Although consumer electronics such as the television have been popular since the 1920s, recycling of them was almost unheard of until early 1991. The first electronic waste recycling scheme was implemented in Switzerland, beginning with collection of old refrigerators but gradually expanding to cover all devices. After these schemes were set up, many countries did not have the capacity to deal with the sheer quantity of e-waste they generated or its hazardous nature. They began to export the problem to developing countries without enforced environmental legislation. This is cheaper, as recycling computer monitors in the United States costs 10 times more than in China. Demand in Asia for electronic waste began to grow when scrap yards found that they could extract valuable substances such as copper, silver, iron, silicon, nickel and gold, during the recycling process. The 2000s saw a large increase in both the sale of electronic devices and their growth as a waste stream: in 2002, e-waste grew faster than any other type of waste in the EU. This caused investment in modern, automated facilities to cope with the influx of redundant appliances, especially after strict laws were implemented in 2003.

As of 2014, the European Union has about 50% of world share of the waste and recycling industries, with over 60,000 companies employing 500,000 persons, with a turnover of €24 billion. Countries have to reach recycling rates of at least 50%, while the lead countries are around 65% and the EU average is 39% as of 2013.

Legislation

Supply

For a recycling program to work, having a large, stable supply of recyclable material is crucial. Three legislative options have been used to create such a supply: mandatory recycling collection, container deposit legislation and refuse bans. Mandatory collection laws set recycling targets for cities to aim for, usually in the form that a certain percentage of a material must be diverted from the city's waste stream by a target date. The city is then responsible for working to meet this target.

Container deposit legislation involves offering a refund for the return of certain containers, typically glass, plastic and metal. When a product in such a container is purchased, a small surcharge is added to the price. This surcharge can be reclaimed by the consumer if the container is returned to a collection point. These programs have been very successful, often resulting in an 80 percent recycling rate. Despite such good results, the shift in collection costs from local government to industry and consumers has created strong opposition to the creation of such programs in some areas. A variation on this is where the manufacturer bears responsibility for the recycling of their goods. In the European Union, the WEEE Directive requires producers of consumer electronics to reimburse the recyclers' costs.

An alternative way to increase supply of recyclates is to ban the disposal of certain materials as waste, often including used oil, old batteries, tires and garden waste. One aim of this method is to create a viable economy for proper disposal of banned products. Care must be taken that enough of these recycling services exist, or such bans simply lead to increased illegal dumping.

Government-mandated Demand

Legislation has also been used to increase and maintain a demand for recycled materials. Four methods of such legislation exist: minimum recycled content mandates, utilization rates, procurement policies and recycled product labeling.

Both minimum recycled content mandates and utilization rates increase demand directly by forcing manufacturers to include recycling in their operations. Content mandates specify that a certain percentage of a new product must consist of recycled material. Utilization rates are a more flexible option: industries are permitted to meet the recycling targets at any point of their operation or even contract recycling out in exchange for tradeable credits. Opponents to both of these methods point to the large increase in reporting requirements they impose, and claim that they rob industry of necessary flexibility.

Governments have used their own purchasing power to increase recycling demand through what are called "procurement policies." These policies are either "set-asides," which reserve a certain amount of spending solely towards recycled products, or "price preference" programs which provide a larger budget when recycled items are purchased. Additional regulations can target specific cases: in the United States, for example, the Environmental Protection Agency mandates the purchase of oil, paper, tires and building insulation from recycled or re-refined sources whenever possible.

The final government regulation towards increased demand is recycled product labeling. When producers are required to label their packaging with amount of recycled material in the product (including the packaging), consumers are better able to make educated choices. Consumers with sufficient buying power can then choose more environmentally conscious options, prompt producers to increase the amount of recycled material in their products, and indirectly increase demand. Standardized recycling labeling can also have a positive effect on supply of recyclates if the labeling includes information on how and where the product can be recycled.

Recyclates

Glass recovered by crushing only one kind of beer bottle

Recyclate is a raw material that is sent to, and processed in a waste recycling plant or materials recovery facility which will be used to form new products. The material is collected in various methods and delivered to a facility where it undergoes re-manufacturing so that it can used in the production of new materials or products. For example, plastic bottles that are collected can be re-used and made into plastic pellets, a new product.

Quality of Recyclate

The quality of recyclates is recognized as one of the principal challenges that needs to be addressed for the success of a long-term vision of a green economy and achieving zero waste. Recyclate quality is generally referring to how much of the raw material is made up of target material compared to the amount of non-target material and other non-recyclable material. Only target material is likely to be recycled, so a higher amount of non-target and non-recyclable material will reduce the quantity of recycling product. A high proportion of non-target and non-recyclable material can make it more difficult for re-processors to achieve "high-quality" recycling. If the recyclate is of poor quality, it is more likely to end up being down-cycled or, in more extreme cases, sent to other recovery options or landfilled. For example, to facilitate the re-manufacturing of clear glass products there are tight restrictions for colored glass going into the re-melt process.

The quality of recyclate not only supports high quality recycling, but it can also deliver significant environmental benefits by reducing, reusing and keeping products out of landfills. High quality recycling can help support growth in the economy by maximizing the economic value of the waste material collected. Higher income levels from the sale of quality recyclates can return value which can be significant to local governments, households and businesses. Pursuing high quality recycling can also provide consumer and business confidence in the waste and resource management sector and may encourage investment in that sector.

There are many actions along the recycling supply chain that can influence and affect the material quality of recyclate. It begins with the waste producers who place non-target and non-recyclable wastes in recycling collection. This can affect the quality of final recyclate streams or require further efforts to discard those materials at later stages in the recycling process. The different collection systems can result in different levels of contamination. Depending on which materials are collected together, extra effort is required to sort this material back into separate streams and can significantly reduce the quality of the final product. Transportation and the compaction of materials can make it more difficult to separate material back into separate waste streams. Sorting facilities are not one hundred per cent effective in separating materials, despite improvements in technology and quality recyclate which can see a loss in recyclate quality. The storage of materials outside where the product can become wet can cause problems for re-processors. Reprocessing facilities may require further sorting steps to further reduce the amount of non-target and non-recyclable material. Each action along the recycling path plays a part in the quality of recyclate.

Quality Recyclate Action Plan (Scotland)

The Recyclate Quality Action Plan of Scotland sets out a number of proposed actions that the Scottish Government would like to take forward in order to drive up the quality of the materials being collected for recycling and sorted at materials recovery facilities before being exported or sold on to the reprocessing market.

The plan's objectives are to:

- Drive up the quality of recyclate.
- Deliver greater transparency about the quality of recyclate.
- Provide help to those contracting with materials recycling facilities to identify what is required of them
- Ensure compliance with the Waste (Scotland) regulations 2012.
- Stimulate a household market for quality recyclate.
- Address and reduce issues surrounding the Waste Shipment Regulations.

The plan focuses on three key areas, with fourteen actions which were identified to increase the quality of materials collected, sorted and presented to the processing market in Scotland.

The three areas of focus are:

1. Collection systems and input contamination
2. Sorting facilities – material sampling and transparency
3. Material quality benchmarking and standards

Recycling Consumer Waste

Collection

A number of different systems have been implemented to collect recyclates from the general waste stream. These systems lie along the spectrum of trade-off between public convenience and government ease and expense. The three main categories of collection are "drop-off centers," "buy-back centers" and "curbside collection."

A three-sided bin at a railway station in Germany, intended to separate paper *(left)* and plastic wrappings *(right)* from other waste *(back)*

Curbside Collection

Curbside collection encompasses many subtly different systems, which differ mostly on where in the process the recyclates are sorted and cleaned. The main categories are mixed waste collection, commingled recyclables and source separation. A waste collection vehicle generally picks up the waste.

At one end of the spectrum is mixed waste collection, in which all recyclates are collected mixed in with the rest of the waste, and the desired material is then sorted out and cleaned at a central sorting facility.

A recycling truck collecting the contents of a recycling bin in Canberra, Australia

This results in a large amount of recyclable waste, paper especially, being too soiled to reprocess, but has advantages as well: the city need not pay for a separate collection of recyclates and no public education is needed. Any changes to which materials are recyclable is easy to accommodate as all sorting happens in a central location.

In a commingled or single-stream system, all recyclables for collection are mixed but kept separate from other waste. This greatly reduces the need for post-collection cleaning but does require public education on what materials are recyclable.

Source separation is the other extreme, where each material is cleaned and sorted prior to collection. This method requires the least post-collection sorting and produces the purest recyclates, but incurs additional operating costs for collection of each separate material. An extensive public education program is also required, which must be successful if recyclate contamination is to be avoided.

Source separation used to be the preferred method due to the high sorting costs incurred by commingled (mixed waste) collection. Advances in sorting technology, however, have lowered this overhead substantially—many areas which had developed source separation programs have since switched to co-mingled collection.

Buy-back Centers

Buy-back centers differ in that the cleaned recyclates are purchased, thus providing a clear incentive for use and creating a stable supply. The post-processed material can then be sold on, hopefully creating a profit. Unfortunately, government subsidies are necessary to make buy-back centres a viable enterprise, as according to the U.S. National Waste & Recycling Association, it costs on average US$50 to process a ton of material, which can only be resold for US$30.

Drop-off Centers

Drop-off centers require the waste producer to carry the recyclates to a central location, either an installed or mobile collection station or the reprocessing plant itself. They are the easiest type of collection to establish, but suffer from low and unpredictable throughput.

Distributed Recycling

For some waste materials such as plastic, recent technical devices called recyclebots enable a form of distributed recycling. Preliminary life-cycle analysis (LCA) indicates that such distributed recycling of HDPE to make filament of 3-D printers in rural regions is energetically favorable to either using virgin resin or conventional recycling processes because of reductions in transportation energy.

Sorting

Once commingled recyclates are collected and delivered to a central collection facility, the different types of materials must be sorted. This is done in a series of stages, many of which involve automated processes such that a truckload of material can be fully sorted in less than an hour. Some plants can now sort the materials automatically, known as single-stream recycling. In plants, a variety of materials are sorted such as paper, different types of plastics, glass, metals, food scraps and most types of batteries. A 30 percent increase in recycling rates has been seen in the areas where these plants exist.

Recycling sorting facility and processes

Initially, the commingled recyclates are removed from the collection vehicle and placed on a conveyor belt spread out in a single layer. Large pieces of corrugated fiberboard and plastic bags are removed by hand at this stage, as they can cause later machinery to jam.

Early sorting of recyclable materials: glass and plastic bottles in Poland

Next, automated machinery such as disk screens and air classifiers separate the recyclates by weight, splitting lighter paper and plastic from heavier glass and metal. Cardboard is removed from the mixed paper and the most common types of plastic, PET (#1) and HDPE (#2), are collected. This separation is usually done by hand but has become automated in some sorting centers: a spectroscopic scanner is used to differentiate between different types of paper and plastic based on the absorbed wavelengths, and subsequently divert each material into the proper collection channel.

Strong magnets are used to separate out ferrous metals, such as iron, steel and tin cans. Non-ferrous metals are ejected by magnetic eddy currents in which a rotating magnetic field induces an electric current around the aluminum cans, which in turn creates a magnetic eddy current inside the cans. This magnetic eddy current is repulsed by a large magnetic field, and the cans are ejected from the rest of the recyclate stream.

A recycling point in New Byth, Scotland, with separate containers for paper, plastics and differently colored glass

Finally, glass is sorted according to its color: brown, amber, green or clear. It may either be sorted by hand, or via an automated machine that uses colored filters to detect different colors. Glass fragments smaller than 10 millimetres (0.39 in) across cannot be sorted automatically, and are mixed together as "glass fines."

This process of recycling as well as reusing the recycled material has proven advantageous because it reduces amount of waste sent to landfills, conserves natural resources, saves energy, reduces greenhouse gas emissions and helps create new jobs. Recycled materials can also be converted into new products that can be consumed again, such as paper, plastic and glass.

The City and County of San Francisco's Department of the Environment is attempting to achieve a city-wide goal of Zero Waste by 2020. San Francisco's refuse hauler, Recology, operates an effective recyclables sorting facility in San Francisco, which helped San Francisco reach a record-breaking diversion rate of 80%.

Rinsing

Food packaging should no longer contain any organic matter (organic matter, if any, needs to be placed in a biodegradable waste bin or be buried in a garden). Since no trace of biodegradable material is best kept in the packaging before placing it in a trash bag, some packaging also needs to be rinsed.

Recycling Industrial Waste

Although many government programs are concentrated on recycling at home, a 64% of waste in the United Kingdom is generated by industry. The focus of many recycling programs done by industry is the cost–effectiveness of recycling. The ubiquitous nature of cardboard packaging makes cardboard a commonly recycled waste product by companies that deal heavily in packaged goods, like retail stores, warehouses and distributors of goods. Other industries deal in niche or specialized products, depending on the nature of the waste materials that are present.

Mounds of shredded rubber tires are ready for processing

The glass, lumber, wood pulp and paper manufacturers all deal directly in commonly recycled materials; however, old rubber tires may be collected and recycled by independent tire dealers for a profit.

Levels of metals recycling are generally low. In 2010, the International Resource Panel, hosted by the United Nations Environment Programme (UNEP) published reports on metal stocks that exist within society and their recycling rates. The Panel reported that the increase in the use of metals during the 20th and into the 21st century has led to a substantial shift in metal stocks from below ground to use in applications within society above ground. For example, the in-use stock of copper in the USA grew from 73 to 238 kg per capita between 1932 and 1999.

The report authors observed that, as metals are inherently recyclable, the metal stocks in society can serve as huge mines above ground (the tW has been coined with this idea in mind). However, they found that the recycling rates of many metals are very low. The report warned that the recycling rates of some rare metals used in applications such as mobile phones, battery packs for hybrid cars and fuel cells, are so low that unless future end-of-life recycling rates are dramatically stepped up these critical metals will become unavailable for use in modern technology.

The military recycles some metals. The U.S. Navy's Ship Disposal Program uses ship breaking to reclaim the steel of old vessels. Ships may also be sunk to create an artificial reef. Uranium is a very dense metal that has qualities superior to lead and titanium for many military and industrial uses. The uranium left over from processing it into nuclear weapons and fuel for nuclear reactors is called depleted uranium, and it is used by all branches of the U.S. military use for armour-piercing shells and shielding.

The construction industry may recycle concrete and old road surface pavement, selling their waste materials for profit.

Some industries, like the renewable energy industry and solar photovoltaic technology in particular, are being proactive in setting up recycling policies even before there is considerable volume to their waste streams, anticipating future demand during their rapid growth.

Aerial photo of a ship recycling facility in Chittagong, Bangladesh

Recycling of plastics is more difficult, as most programs are not able to reach the necessary level of quality. Recycling of PVC often results in downcycling of the material, which means only products of lower quality standard can be made with the recycled material. A new approach which allows an equal level of quality is the Vinyloop process. It was used after the London Olympics 2012 to fulfill the PVC Policy.

E-waste Recycling

E-waste is a growing problem, accounting for 20-50 million metric tons of global waste per year according to the EPA. It is also the fastest growing waste stream in the EU. Many recyclers do not recycle e-waste responsibly. After the cargo barge Khian Sea dumped 14,000 metric tons of toxic ash in Haiti, the Basel Convention was formed to stem the flow of hazardous substances into poorer countries. They created the e-Stewards certification to ensure that recyclers are held to the highest standards for environmental responsibility and to help consumers identify responsible recyclers. This works alongside other prominent legislation, such as the Waste Electrical and Electronic Equipment Directive of the EU the United States National Computer Recycling Act, to prevent poisonous chemicals from entering waterways and the atmosphere.

Microprocessors retrieved from waste stream

In the recycling process, television sets, monitors, cell phones and computers are typically tested for reuse and repaired. If broken, they may be disassembled for parts still having high value if labor is cheap enough. Other e-waste is shredded to pieces roughly 10 centimetres (3.9 in) in size, and manually checked to separate out toxic batteries and capacitors which contain poisonous metals. The remaining pieces are further shredded to 10 millimetres (0.39 in) particles and passed under a magnet to remove ferrous metals. An eddy current ejects non-ferrous metals, which are sorted by density either by a centrifuge or vibrating plates. Precious metals can be dissolved in acid, sorted, and smelted into ingots. The remaining glass and plastic fractions are separated by density and sold to re-processors. Television sets and monitors must be manually disassembled to remove lead from CRTs or the mercury backlight from LCDs.

Plastic Recycling

A container for recycling used plastic spoons into material for 3D printing

Plastic recycling is the process of recovering scrap or waste plastic and reprocessing the material into useful products, sometimes completely different in form from their original state. For instance, this could mean melting down soft drink bottles and then casting them as plastic chairs and tables.

Physical Recycling

Some plastics are remelted to form new plastic objects; for example, PET water bottles can be converted into polyester destined for clothing. A disadvantage of this type of recycling is that the molecular weight of the polymer can change further and the levels of unwanted substances in the plastic can increase with each remelt.

Chemical Recycling

For some polymers, it is possible to convert them back into monomers, for example PET can be treated with an alcohol and a catalyst to form a dialkyl terephthalate. The terephthalate diester can be used with ethylene glycol to form a new polyester polymer, thus making it possible to use the pure polymer again.

Waste Plastic Pyrolysis to Fuel Oil

Another process involves conversion of assorted polymers into petroleum by a much less precise thermal depolymerization process. Such a process would be able to accept almost any polymer or mix of polymers, including thermoset materials such as vulcanized rubber tires and the biopolymers in feathers and other agricultural waste. Like natural petroleum, the chemicals produced can be used as fuels or as feedstock. A RESEM Technology plant of this type in Carthage, Missouri, USA, uses turkey waste as input material. Gasification is a similar process, but is not technically recycling since polymers are not likely to become the result. Plastic Pyrolysis can convert petroleum based waste streams such as plastics into quality fuels, carbons. Given below is the list of suitable plastic raw materials for pyrolysis:

- Mixed plastic (HDPE, LDPE, PE, PP, Nylon, Teflon, PS, ABS, FRP, etc.)

- Mixed waste plastic from waste paper mill

- Multi-layered plastic

Recycling Codes

In order to meet recyclers' needs while providing manufacturers a consistent, uniform system, a coding system was developed. The recycling code for plastics was introduced in 1988 by the plastics industry through the Society of the Plastics Industry. Because municipal recycling programs traditionally have targeted packaging—primarily bottles and containers—the resin coding system offered a means of identifying the resin content of bottles and containers commonly found in the residential waste stream.

Plastic products are printed with numbers 1–7 depending on the type of resin. Type 1 (polyethylene terephthalate) is commonly found in soft drink and water bottles. Type 2 (high-density polyethylene) is found in most hard plastics such as milk jugs, laundry detergent bottles and some dishware. Type 3 (polyvinyl chloride) includes items such as shampoo bottles, shower curtains, hoola hoops, credit cards, wire jacketing, medical equipment, siding and piping. Type 4 (low-density polyethylene) is found in shopping bags, squeezable bottles, tote bags, clothing, furniture and carpet. Type 5 is polypropylene and makes up syrup bottles, straws, Tupperware and some automotive parts. Type 6 is polystyrene and makes up meat trays, egg cartons, clamshell containers and compact disc cases. Type 7 includes all other plastics such as bulletproof materials, 3- and 5-gallon water bottles and sunglasses. Having a recycling code or the chasing arrows logo on a material is not an automatic indicator that a material is recyclable but rather an explanation of what the material is. Types 1 and 2 are the most commonly recycled.

Recycling codes on products

Economic Impact

Critics dispute the net economic and environmental benefits of recycling over its costs, and suggest that proponents of recycling often make matters worse and suffer from confirmation bias. Specifically, critics argue that the costs and energy used in collection and transportation detract from (and outweigh) the costs and energy saved in the production process; also that the jobs produced by the recycling industry can be a poor trade for the jobs lost in logging, mining, and other industries associated with production; and that materials such as paper pulp can only be recycled a few times before material degradation prevents further recycling.

The National Waste and Recycling Association (NWRA), reported in May 2015, that recycling and waste made a $6.7 billion economic impact in Ohio, U.S., and employed 14,000 people.

Cost–benefit Analysis

Environmental effects of recycling		
Material	**Energy savings**	**Air pollution savings**
Aluminium	95%	95%
Cardboard	24%	—
Glass	5–30%	20%
Paper	40%	73%
Plastics	70%	—
Steel	60%	—

There is some debate over whether recycling is economically efficient. It is said that dumping 10,000 tons of waste in a landfill creates six jobs, while recycling 10,000 tons of waste can create over 36 jobs. However, the cost effectiveness of creating the additional jobs remains unproven. According to the U.S. Recycling Economic Informational Study, there are over 50,000 recycling establishments that have created over a million jobs in the US. Two years after New York City declared that implementing recycling programs would be "a drain on the city," New York City leaders realized that an efficient recycling system could save the city over $20 million. Municipalities often see fiscal benefits from implementing recycling programs, largely due to the reduced landfill costs. A study conducted by the Technical University of Denmark according to the Economist found that in 83 percent of cases, recycling is the most efficient method to dispose of household waste. However, a 2004 assessment by the Danish Environmental Assessment Institute concluded that incineration was the most effective method for disposing of drink containers, even aluminium ones.

Fiscal efficiency is separate from economic efficiency. Economic analysis of recycling does not include what economists call externalities, which are unpriced costs and benefits that accrue to individuals outside of private transactions. Examples include: decreased air pollution and greenhouse gases from incineration, reduced hazardous waste leaching from landfills, reduced energy consumption, and reduced waste and resource consumption, which leads to a reduction in environmentally damaging mining and timber activity. About 4,000 minerals are known, of these only a few hundred minerals in the world are relatively common. Known reserves of phosphorus will be exhausted within the next 100 years at current rates of usage. Without mechanisms such as taxes or subsidies to internalize externalities, businesses will ignore them despite the costs imposed on society. To make such nonfiscal benefits economically relevant, advocates have pushed for legislative action to increase the demand for recycled materials. The United States Environmental Protection Agency (EPA) has concluded in favor of recycling, saying that recycling efforts reduced the country's carbon emissions by a net 49 million metric tonnes in 2005. In the United Kingdom, the Waste and Resources Action Programme stated that Great Britain's recycling efforts reduce CO_2 emissions by 10–15 million tonnes a year. Recycling is more efficient in densely populated areas, as there are economies of scale involved.

Wrecked automobiles gathered for smelting

Certain requirements must be met for recycling to be economically feasible and environmentally effective. These include an adequate source of recyclates, a system to extract those recyclates from the waste stream, a nearby factory capable of reprocessing the recyclates, and a potential demand for the recycled products. These last two requirements are often overlooked—without both an industrial market for production using the collected materials and a consumer market for the manufactured goods, recycling is incomplete and in fact only "collection".

Free-market economist Julian Simon remarked "There are three ways society can organize waste disposal: (a) commanding,(b) guiding by tax and subsidy, and (c) leaving it to the individual and the market". These principles appear to divide economic thinkers today.

Frank Ackerman favours a high level of government intervention to provide recycling services. He believes that recycling's benefit cannot be effectively quantified by traditional *laissez-faire* economics. Allen Hershkowitz supports intervention, saying that it is a public service equal to education and policing. He argues that manufacturers should shoulder more of the burden of waste disposal.

Paul Calcott and Margaret Walls advocate the second option. A deposit refund scheme and a small refuse charge would encourage recycling but not at the expense of fly-tipping. Thomas C. Kinnaman concludes that a landfill tax would force consumers, companies and councils to recycle more.

Most free-market thinkers detest subsidy and intervention because they waste resources. Terry Anderson and Donald Leal think that all recycling programmes should be privately operated, and therefore would only operate if the money saved by recycling exceeds its costs. Daniel K. Benjamin argues that it wastes people's resources and lowers the wealth of a population.

Trade in Recyclates

Certain countries trade in unprocessed recyclates. Some have complained that the ultimate fate of recyclates sold to another country is unknown and they may end up in landfills instead of reprocessed. According to one report, in America, 50–80 percent of computers destined for recycling are actually not recycled. There are reports of illegal-waste imports to China being dismantled and recycled solely for monetary gain, without consideration for workers' health or environmental damage. Although the Chinese government has banned these practices, it has not been able to eradicate them. In 2008, the prices of recyclable waste plummeted before rebounding in 2009. Cardboard averaged about £53/tonne from 2004–2008, dropped to £19/tonne, and then went up to £59/tonne in May 2009. PET plastic averaged about £156/tonne, dropped to £75/tonne and then moved up to £195/tonne in May 2009.

Certain regions have difficulty using or exporting as much of a material as they recycle. This problem is most prevalent with glass: both Britain and the U.S. import large quantities of wine bottled in green glass. Though much of this glass is sent to be recycled, outside the American Midwest there is not enough wine production to use all of the reprocessed material. The extra must be downcycled into building materials or re-inserted into the regular waste stream.

Similarly, the northwestern United States has difficulty finding markets for recycled newspaper, given the large number of pulp mills in the region as well as the proximity to Asian markets. In other areas of the U.S., however, demand for used newsprint has seen wide fluctuation.

In some U.S. states, a program called RecycleBank pays people to recycle, receiving money from local municipalities for the reduction in landfill space which must be purchased. It uses a single stream process in which all material is automatically sorted.

Criticisms and Responses

Much of the difficulty inherent in recycling comes from the fact that most products are not designed with recycling in mind. The concept of sustainable design aims to solve this problem, and was laid out in the book *Cradle to Cradle: Remaking the Way We Make Things* by architect William McDonough and chemist Michael Braungart. They suggest that every product (and all packaging they require) should have a complete "closed-loop" cycle mapped out for each component—a way in which every component will either return to the natural ecosystem through biodegradation or be recycled indefinitely.

Complete recycling is impossible from a practical standpoint. In summary, substitution and recycling strategies only delay the depletion of non-renewable stocks and therefore may buy time in the transition to true or strong sustainability, which ultimately is only guaranteed in an economy based on renewable resources.

—M. H. Huesemann, 2003

While recycling diverts waste from entering directly into landfill sites, current recycling misses the dissipative components. Complete recycling is impracticable as highly dispersed wastes become so diluted that the energy needed for their recovery becomes increasingly excessive. "For example, how will it ever be possible to recycle the numerous chlorinated organic hydrocarbons that have bioaccumulated in animal and human tissues across the globe, the copper dispersed in fungicides, the lead in widely applied paints, or the zinc oxides present in the finely dispersed rubber powder that is abraded from automobile tires?"

As with environmental economics, care must be taken to ensure a complete view of the costs and benefits involved. For example, paperboard packaging for food products is more easily recycled than most plastic, but is heavier to ship and may result in more waste from spoilage.

Energy and Material Flows

The amount of energy saved through recycling depends upon the material being recycled and the type of energy accounting that is used. Correct accounting for this saved energy can be accomplished with life-cycle analysis using real energy values. In addition, exergy, which is a measure of useful energy can be used. In general, it takes far less energy to produce a unit mass of recycled materials than it does to make the same mass of virgin materials.

Bales of crushed steel ready for transport to the smelter

Some scholars use emergy (spelled with an m) analysis, for example, budgets for the amount of energy of one kind (exergy) that is required to make or transform things into another kind of product or service. Emergy calculations take into account economics which can alter pure physics based results. Using emergy life-cycle analysis researchers have concluded that materials with large refining costs have the greatest potential for high recycle benefits. Moreover, the highest emergy efficiency accrues from systems geared toward material recycling, where materials are engineered to recycle back into their original form and purpose, followed by adaptive reuse systems where the materials are recycled into a different kind of product, and then by-product reuse systems where parts of the products are used to make an entirely different product.

The Energy Information Administration (EIA) states on its website that "a paper mill uses 40 percent less energy to make paper from recycled paper than it does to make paper from fresh lumber." Some critics argue that it takes more energy to produce recycled products than it does to dispose of them in traditional landfill methods, since the curbside collection of recyclables often requires a second waste truck. However, recycling proponents point out that a second timber or logging truck is eliminated when paper is collected for recycling, so the net energy consumption is the same. An Emergy life-cycle analysis on recycling revealed that fly ash, aluminum, recycled concrete aggregate, recycled plastic, and steel yield higher efficiency ratios, whereas the recycling of lumber generates the lowest recycle benefit ratio. Hence, the specific nature of the recycling process, the methods used to analyse the process, and the products involved affect the energy savings budgets.

It is difficult to determine the amount of energy consumed or produced in waste disposal processes in broader ecological terms, where causal relations dissipate into complex networks of material and energy flow. For example, "cities do not follow all the strategies of ecosystem development. Biogeochemical paths become fairly straight relative to wild ecosystems, with very reduced recycling, resulting in large flows of waste and low total energy efficiencies. By contrast, in wild ecosystems, one population's wastes are another population's resources, and succession results in efficient exploitation of available resources. However, even modernized cities may still be in the earliest stages of a succession that may take centuries or millennia to complete." How much energy is used in recycling also depends on the type of material being recycled and the process used to do so. Aluminium is generally agreed to use far less energy when recycled rather than being produced from scratch. The EPA states that "recycling aluminum cans, for example, saves 95 percent of the energy required to make the same amount of aluminum from its virgin source, bauxite." In 2009 more than half of all aluminium cans produced came from recycled aluminium.

Every year, millions of tons of materials are being exploited from the earth's crust, and processed into consumer and capital goods. After decades to centuries, most of these materials are "lost". With the exception of some pieces of art or religious relics, they are no longer engaged in the consumption process. Where are they? Recycling is only an intermediate solution for such materials, although it does prolong the residence time in the anthroposphere. For thermodynamic reasons, however, recycling cannot prevent the final need for an ultimate sink.

—P. H. Brunner

Economist Steven Landsburg has suggested that the sole benefit of reducing landfill space is trumped by the energy needed and resulting pollution from the recycling process. Others, however, have calculated through life-cycle assessment that producing recycled paper uses less energy and water than harvesting, pulping, processing, and transporting virgin trees. When less recycled paper is used, additional energy is needed to create and maintain farmed forests until these forests are as self-sustainable as virgin forests.

Other studies have shown that recycling in itself is inefficient to perform the "decoupling" of economic development from the depletion of non-renewable raw materials that is necessary for sustainable development. The international transportation or recycle material flows through "... different trade networks of the three countries result in different flows, decay rates, and potential recycling returns." As global consumption of a natural resources grows, its depletion is inevitable. The best recycling can do is to delay, complete closure of material loops to achieve 100 percent recycling of nonrenewables is impossible as micro-trace materials dissipate into the environment causing severe damage to the planet's ecosystems. Historically, this was identified as the metabolic rift by Karl Marx, who identified the unequal exchange rate between energy and nutrients flowing from rural areas to feed urban cities that create effluent wastes degrading the planet's ecological capital, such as loss in soil nutrient production. Energy conservation also leads to what is known as Jevon's paradox, where improvements in energy efficiency lowers the cost of production and leads to a rebound effect where rates of consumption and economic growth increases.

A shop in New York only sells items recycled from demolished buildings

Costs

The amount of money actually saved through recycling depends on the efficiency of the recycling program used to do it. The Institute for Local Self-Reliance argues that the cost of recycling depends on various factors, such as landfill fees and the amount of disposal that the community recycles. It states that communities start to save money when they treat recycling as a replacement for their traditional waste system rather than an add-on to it and by "redesigning their collection schedules and/or trucks."

In some cases, the cost of recyclable materials also exceeds the cost of raw materials. Virgin plastic resin costs 40 percent less than recycled resin. Additionally, a United States Environmental Protection Agency (EPA) study that tracked the price of clear glass from July 15 to August 2, 1991, found that the average cost per ton ranged from $40 to $60, while a USGS report shows that the cost per ton of raw silica sand from years 1993 to 1997 fell between $17.33 and $18.10.

In 1996 and 2015 articles for *The New York Times*, John Tierney argued that it costs more money to recycle the trash of New York City than it does to dispose of it in a landfill. Tierney argued that the recycling process employs people to do the additional waste disposal, sorting, inspecting, and many fees are often charged because the processing costs used to make the end product are often more than the profit from its sale. Tierney also referenced a study conducted by the Solid Waste Association of North America (SWANA) that found in the six communities involved in the study, "all but one of the curbside recycling programs, and all the composting operations and waste-to-energy incinerators, increased the cost of waste disposal."

Tierney also points out that "the prices paid for scrap materials are a measure of their environmental value as recyclables. Scrap aluminum fetches a high price because recycling it consumes so much less energy than manufacturing new aluminum."

However, comparing the market cost of recyclable material with the cost of new raw materials ignores economic externalities—the costs that are currently not counted by the market. Creating a new piece of plastic, for instance, may cause more pollution and be less sustainable than recycling a similar piece of plastic, but these factors will not be counted in market cost. A life cycle assessment can be used to determine the levels of externalities and decide whether the recycling may be worthwhile despite unfavorable market costs. Alternatively, legal means (such as a carbon tax) can be used to bring externalities into the market, so that the market cost of the material becomes close to the true cost.

In a 2007 article, Michael Munger, chairman of political science at Duke University, wrote that "if recycling is more expensive than using new materials, it can't possibly be efficient.... There is a simple test for determining whether something is a resource... or just garbage... If someone will pay you for the item, it's a resource.... But if you have to pay someone to take the item away,... then the item is garbage."

In a 2002 article for The Heartland Institute, Jerry Taylor, director of natural resource studies at the Cato Institute, wrote, "If it costs X to deliver newly manufactured plastic to the market, for example, but it costs 10X to deliver reused plastic to the market, we can conclude the resources required to recycle plastic are 10 times more scarce than the resources required to make plastic from scratch. And because recycling is supposed to be about the conservation of resources, mandating recycling under those circumstances will do more harm than good."

Working Conditions

People in Brazil who earn their living by collecting and sorting garbage and selling them for recycling

The recycling of waste electrical and electronic equipment in India and China generates a significant amount of pollution. Informal recycling in an underground economy of these countries has generated an environmental and health disaster. High levels of lead (Pb), polybrominated diphenylethers (PBDEs), polychlorinated dioxins and furans, as well as polybrominated dioxins and furans (PCDD/Fs and PBDD/Fs) concentrated in the air, bottom ash, dust, soil, water and sediments in areas surrounding recycling

Because the social support of a country is likely to be less than the loss of income to the poor undertaking recycling, there is a greater chance the poor will come in conflict with the large recycling organizations. This means fewer people can decide if certain waste is more economically reusable in its current form rather than being reprocessed. Contrasted to the recycling poor, the efficiency of their recycling may actually be higher for some materials because individuals have greater control over what is considered "waste."

One labor-intensive underused waste is electronic and computer waste. Because this waste may still be functional and wanted mostly by those on lower incomes, who may sell or use it at a greater efficiency than large recyclers.

Some recycling advocates believe that laissez-faire individual-based recycling does not cover all of society's recycling needs. Thus, it does not negate the need for an organized recycling program. Local government can consider the activities of the recycling poor as contributing to property blight.

Public Participation Rates

Changes that have been demonstrated to increase recycling rates include:

- Single-stream recycling

- Pay as you throw fees for trash

"Between 1960 and 2000, the world production of plastic resins increased 25-fold, while recovery of the material remained below 5 percent." Many studies have addressed recycling behaviour and strategies to encourage community involvement in recycling programmes. It has been argued that recycling behaviour is not natural because it requires a focus and appreciation for long-term planning, whereas humans have evolved to be sensitive to short-term survival goals; and that to overcome this innate predisposition, the best solution would be to use social pressure to compel participation in recycling programmes. However, recent studies have concluded that social pressure is unviable in this context. One reason for this is that social pressure functions well in small group sizes of 50 to 150 individuals (common to nomadic hunter–gatherer peoples) but not in communities numbering in the millions, as we see today. Another reason is that individual recycling does not take place in the public view.

In a study done by social psychologist Shawn Burn, it was found that personal contact with individuals within a neighborhood is the most effective way to increase recycling within a community. In his study, he had 10 block leaders talk to their neighbors and persuade them to recycle. A comparison group was sent fliers promoting recycling. It was found that the neighbors that were personally contacted by their block leaders recycled much more than the group without personal contact. As a result of this study, Shawn Burn believes that personal contact within a small group of people is an important factor in encouraging recycling. Another study done by Stuart Oskamp examines the effect of neighbors and friends on recycling. It was found in his studies that people who had friends and neighbors that recycled were much more likely to also recycle than those who didn't have friends and neighbors that recycled.

Many schools have created recycling awareness clubs in order to give young students an insight on recycling. These schools believe that the clubs actually encourage students to not only recycle at school, but at home as well.

References

- Cleveland, Cutler J.; Morris, Christopher G. (November 15, 2013). Handbook of Energy: Chronologies, Top Ten Lists, and Word Clouds. Elsevier. p. 461. ISBN 978-0-12-417019-3.

- Carl A. Zimring (2005). Cash for Your Trash: Scrap Recycling in America. New Brunswick, NJ: Rutgers University Press. ISBN 0-8135-4694-X.

- Lynn R. Kahle; Eda Gurel-Atay, eds. (2014). Communicating Sustainability for the Green Economy. New York: M.E. Sharpe. ISBN 978-0-7656-3680-5.

- Sahni, S.; Gutowski, T. G. (2011). "Your scrap, my scrap! The flow of scrap materials through international trade" (PDF). IEEE International Symposium on Sustainable Systems and Technology (ISSST): 1–6. doi:10.1109/ISSST.2011.5936853. ISBN 978-1-61284-394-0.

- Foster, J. B.; Clark, B. (2011). The Ecological Rift: Capitalisms War on the Earth. Monthly Review Press. p. 544. ISBN 1-58367-218-4.

- "Bulgaria opens largest WEEE recycling factory in Eastern Europe". www.ask-eu.com. WtERT Germany GmbH. July 12, 2010. Retrieved July 29, 2015.

- Goodman, Peter S. (January 11, 2012). "Where Gadgets Go To Die: E-Waste Recycler Opens New Plant In Las Vegas". The Huffington Post. Retrieved July 29, 2015.

- RecyclingToday (May 14, 2015). "Recycling and waste have $6.7 billion economic impact in Ohio". Archived from the original on May 18, 2015.

- "Puzzled About Recycling's Value? Look Beyond the Bin" (PDF). United States Environmental Protection Agency. January 1998. Archived from the original (PDF) on June 16, 2015. Retrieved July 31, 2015.

- Pleasant, Barbara (September 26, 2013). "Trench Composting Your Kitchen Waste". www.growveg.com. Growing Interactive Ltd. Retrieved July 31, 2015.

- Hogye, Thomas Q. "The Anatomy of a Computer Recycling Process" (PDF). California Department of Resources Recycling and Recovery. Retrieved October 13, 2014.

- "PM's advisor hails recycling as climate change action.". Letsrecycle.com. November 8, 2006. Archived from the original on August 11, 2007. Retrieved April 15, 2014.

- "Mayor Lee Announces San Francisco Reaches 80 Percent Landfill Waste Diversion, Leads All Cities in North America | sfenvironment.org – Our Home. Our City. Our Planet". sfenvironment.org. October 5, 2012. Retrieved June 9, 2014.

- "ITP Aluminum: Energy and Environmental Profile of the U.S. Aluminum Industry" (PDF). Archived from the original (PDF) on August 11, 2011. Retrieved November 6, 2012.

- "Pratarelli, M.E. (2010) "Social pressure and recycling: a brief review, commentary and extensions". S.A.P.I.EN.S. 3 (1)". Sapiens.revues.org. Retrieved November 6, 2012.

Various Materials Recycling Methods

This chapter divides recycling methods in three broad categories- computer recycling, retrocomputing and greening. Computer recycling techniques detailed in the chapter include data shredder, data erasure, degaussing, data remanence, electronic waste and file deletion. Retrocomputing techniques like home computer remake, minimalism (computing) and abandonware have been discussed. Greening includes methods like community greens, green alliance, community gardening, village green, park and common land. All these methods and their contribution to the recycling efforts have been comprehensively studied in the chapter.

Computer Recycling

Computer recycling, electronic recycling or e-waste recycling is the disassembly and separation of components and raw materials of waste electronics. Although the procedures of re-use, donation and repair are not strictly recycling, they are other common sustainable ways to dispose of IT waste.

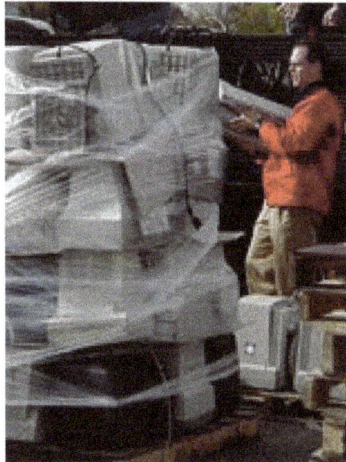

Computer monitors are typically packed into low stacks on wooden pallets
for recycling and then shrink-wrapped.

In 2009, 38% of computers and a quarter of total electronic waste was recycled in the United States, 5% and 3% up from 3 years prior respectively. Since its inception in the early 1990s, more and more devices are recycled worldwide due to increased awareness and investment. Electronic recycling occurs primarily in order to recover valuable rare earth metals and precious metals, which are in short supply, as well as plastics and metals. These are resold or used in new devices after purification, in effect creating a circular economy.

Recycling is considered environmentally friendly because it prevents hazardous waste, including heavy metals and carcinogens, from entering the atmosphere, landfill or waterways. While electronics consist a small fraction of total waste generated, they are far more dangerous. There is stringent legislation designed to enforce and encourage the sustainable disposal of appliances, the most notable being the Waste Electrical and Electronic Equipment Directive of the European Union and the United States National Computer Recycling Act.

Opponents argue that recycling is expensive and ineffective, that it does not safeguard data and that it stifles innovation. It is also criticised for exporting, often illegally, large volumes of toxic waste to countries such as India, China and Nigeria for crude manual disassembly by workers who have little regard for the risk to themselves or the environment.

Reasons for Recycling

Obsolete computers and old electronics are valuable sources for secondary raw materials if recycled; otherwise, these devices are a source of toxins and carcinogens. Rapid technology change, low initial cost, and planned obsolescence have resulted in a fast-growing surplus of computers and other electronic components around the globe. Technical solutions are available, but in most cases a legal framework, collection system, logistics, and other services need to be implemented before applying a technical solution. The U.S. Environmental Protection Agency, estimates 30 to 40 million surplus PCs, classified as "hazardous household waste", would be ready for end-of-life management in the next few years. The U.S. National Safety Council estimates that 75% of all personal computers ever sold are now surplus electronics.

In 2007, the United States Environmental Protection Agency (EPA) stated that more than 63 million computers in the U.S. were traded in for replacements or discarded. Today, 15% of electronic devices and equipment are recycled in the United States. Most electronic waste is sent to landfills or incinerated, which releases materials such as lead, mercury, or cadmium into the soil, groundwater, and atmosphere, thus having a negative impact on the environment.

Many materials used in computer hardware can be recovered by recycling for use in future production. Reuse of tin, silicon, iron, aluminium, and a variety of plastics that are present in bulk in computers or other electronics can reduce the costs of constructing new systems. Components frequently contain lead, copper, gold and other valuable materials suitable for reclamation.

Computer components contain many toxic substances, like dioxins, polychlorinated biphenyls (PCBs), cadmium, chromium, radioactive isotopes and mercury. A typical computer monitor may contain more than 6% lead by weight, much of which is in the lead glass of the cathode ray tube (CRT). A typical 15 inch (38 cm) computer monitor may contain 1.5 pounds (1 kg) of lead but other monitors have been estimated to have up to 8 pounds (4 kg) of lead. Circuit boards contain considerable quantities of lead-tin solders that are more likely to leach into groundwater or create air pollution due to incineration. The processing (e.g. incineration and acid treatments) required to reclaim these precious substances may release, generate, or synthesize toxic byproducts.

Export of waste to countries with lower environmental standards is a major concern. The Basel Convention includes hazardous wastes such as, but not limited to, CRT screens as an item that may not be exported transcontinentally without prior consent of both the country exporting and receiving the waste. Companies may find it cost-effective in the short term to sell outdated computers to less developed countries with lax regulations. It is commonly believed that a majority of surplus laptops are routed to developing nations as "dumping grounds for e-waste". The high value of working and reusable laptops, computers, and components (e.g. RAM) can help pay the cost of transportation for many worthless "commodities". The laws governing the exportation of waste electronics are put in place to stop "recycling companies" in developed countries from shipping their waste to 3rd world countries as working devices; they are never working devices. The 3rd world workers scavenge specific items with selling value and throw the rest away to rot and become a health hazard in their own backyard.

Regulations

An abandoned Taxan monitor.

Europe

In Switzerland, the first electronic waste recycling system was implemented in 1991, beginning with collection of old refrigerators; over the years, all other electric and electronic devices were gradually added to the system. The established producer responsibility organization is SWICO, mainly handling information, communication, and organization technology. The European Union implemented a similar system in February 2003, under the Waste Electrical and Electronic Equipment Directive (WEEE Directive, 2002/96/EC).

Pan European adoption of the Legislation was slow on take-up, with Italy and the United Kingdom being the final member states to pass it into law. The success of the WEEE directive has varied significantly from state to state, with collection rates varying between 13 kilograms per capita per annum to as little as 1 kg per capita per annum. Computers & electronic wastes collected from households within Europe are treated under the WEEE directive via Producer Compliance Schemes (whereby manufacturers of Electronics pay into a scheme that funds its recovery from household waste recycling centres (HWRCs)) and nominated Waste Treatment Facilities (known as Obligated WEEE).

However, recycling of ex corporate Computer Hardware and associated electronic equipment falls outside the Producer Compliance Scheme (Known as non-obligated). In the UK, Waste or obsolete corporate related computer hardware is treated via third party Authorized Treatment Facilities, who normally impose a charge for its collection and treatment.

United States

Federal

The United States Congress considers a number of electronic waste bills, like the National Computer Recycling Act introduced by Congressman Mike Thompson (D-CA). The main federal law governing solid waste is the Resource Conservation and Recovery Act of 1976. It covers only CRTs, though state regulations may differ. There are also separate laws concerning battery disposal. On March 25, 2009, the House Science and Technology Committee approved funding for research on reducing electronic waste and mitigating environmental impact, regarded by sponsor Ralph Hall (R-TX) as the first federal bill to directly address electronic waste.

State

Many states have introduced legislation concerning recycling and reuse of computers or computer parts or other electronics. Most American computer recycling legislations address it from within the larger electronic waste issue.

In 2001, Arkansas enacted the Arkansas Computer and Electronic Solid Waste Management Act, which requires that state agencies manage and sell surplus computer equipment, establishes a computer and electronics recycling fund, and authorizes the Department of Environmental Quality to regulate and/or ban the disposal of computer and electronic equipment in Arkansas landfills.

The recently passed Electronic Device Recycling Research and Development Act distributes grants to universities, government labs and private industries for research in developing projects in line with e-waste recycling and refurbishment.

Asia

In Japan, sellers and manufacturers of certain electronics (such as televisions and air conditioners) are required to recycle them. However, no legislation exists to cover the recycling of computer or cellphone related wastes.

It is required in South Korea and Taiwan that sellers and manufacturers of electronics be responsible for recycling 75% of their used products.

According to a report by UNEP titled, "Recycling - from E-Waste to Resources," the amount of e-waste being produced - including mobile phones and computers - could rise by as much as 500 percent over the next decade in some countries, such as India.

Electronic waste is often exported to developing countries.

4.5-volt, D, C, AA, AAA, AAAA, A23, 9-volt, CR2032 and LR44 cells are all recyclable in most countries.

One theory is that increased regulation of electronic waste and concern over the environmental harm in mature economies creates an economic disincentive to remove residues prior to export. Critics of trade in used electronics maintain that it is too easy for brokers calling themselves recyclers to export unscreened electronic waste to developing countries, such as China, India and parts of Africa, thus avoiding the expense of removing items like bad cathode ray tubes (the processing of which is expensive and difficult). The developing countries are becoming big dump yards of e-waste. Proponents of international trade point to the success of fair trade programs in other industries, where cooperation has led creation of sustainable jobs, and can bring affordable technology in countries where repair and reuse rates are higher.

Organizations like A2Z Group (Company Website) have stepped in to own up the responsibility to collect and recycle e-waste at various locations in India.

South Africa

Thanks to the National Environmental Management Act 1998 and National Environmental Management Waste Act 2008, any person in any position causing harm to the environment and failing to comply with the Waste Act could be fined R10 Million or put into jail or receive both penalties for their transgressions.

Recycling Tethods

Computers being collected for recycling at a pickup event in Olympia, Washington, United States.

Consumer Recycling

Consumer recycling options consists of sale, donating computers directly to organizations in need, sending devices directly back to their original manufacturers, or getting components to a convenient recycler or refurbisher.

Scrapping/Recycling

The rising price of precious metals — coupled with the high rate of unemployment during the Great Recession — has led to a larger number of amateur "for profit" electronics recyclers. Computer parts, for example, are stripped of their most valuable components and sold for scrap. Metals like copper, aluminum, lead, gold and palladium are recovered from computers, televisions and more.

In the recycling process, TVs, monitors, mobile phones and computers are typically tested for reuse and repaired. If broken, they may be disassembled for parts still having high value if labour is cheap enough. Other e-waste is shredded to roughly 100 mm pieces and manually checked to separate out toxic batteries and capacitors which contain poisonous metals. The remaining pieces are further shredded to ~10 mm and passed under a magnet to remove ferrous metals. An eddy current ejects non-ferrous metals, which are sorted by density either by a centrifuge or vibrating plates. Precious metals can be dissolved in acid, sorted, and smelted into ingots. The remaining glass and plastic fractions are separated by density and sold to re-processors. TVs and monitors must be manually disassembled to remove either toxic lead in CRTs or the mercury in flat screens.

Bulk laptops at a recycling affiliate, broken down into Dell, Gateway
Computers, Hewlett-Packard, Sony, and other.

Corporate Recycling

Businesses seeking a cost-effective way to recycle large amounts of computer equipment responsibly face a more complicated process.

Businesses also have the options of sale or contacting the Original Equipment Manufacturers (OEMs) and arranging recycling options.

Some companies pick up unwanted equipment from businesses, wipe the data clean from the systems, and provide an estimate of the product's remaining value. For unwanted items that still have value, these firms buy the excess IT hardware and sell refurbished products to those seeking more affordable options than buying new.

Companies that specialize in data protection and green disposal processes dispose of both data and used equipment, while employing strict procedures to help improve the environment. Professional IT Asset Disposition (ITAD) firms specialize in corporate computer disposal and recycling services in compliance with local laws and regulations and also offer secure data elimination services that comply with Data remanence standards including National Institute of Standards and Technology.

Corporations face risks both for incompletely destroyed data and for improperly disposed computers. In the UK, some recycling companies use a specialized WEEE-registered contractor to dispose IT equipment and electrical appliances, who disposes it safely and legally. In America, companies are liable for compliance with regulations even if the recycling process is outsourced under the Resource Conservation and Recovery Act. Companies can mitigate these risks by requiring waivers of liability, audit trails, certificates of data destruction, signed confidentiality agreements, and random audits of information security. The National Association of Information Destruction is an international trade association for data destruction providers.

Sale

Online auctions are an alternative for consumers willing to resell for cash less fees, in a complicated, self-managed, competitive environment where paid listings might not sell. Online classified ads can be similarly risky due to forgery scams and uncertainty.

Take Back

When researching computer companies before a computer purchase, consumers can find out if they offer recycling services. Most major computer manufacturers offer some form of recycling. At the user's request they may mail in their old computers, or arrange for pickup from the manufacturer.

Hewlett-Packard also offers free recycling, but only one of its "national" recycling programs is available nationally, rather than in one or two specific states. Hewlett-Packard also offers to pick up any computer product of any brand for a fee, and to offer a coupon against the purchase of future computers or components; it was the largest computer recycler in America in 2003, and it has recycled over 750,000,000 pounds (340,000,000 kg) of electronic waste globally since 1995. It encourages the shared approach of collection points for consumers and recyclers to meet.

Exchange

Manufacturers often offer a free replacement service when purchasing a new PC. Dell Computers and Apple Inc. take back old products when one buys a new one. Both refurbish and resell their own computers with a one-year warranty.

Many companies purchase and recycle all brands of working and broken laptops and notebook computers from individuals and corporations. Building a market for recycling of desktop computers has proven more difficult than exchange programs for laptops, smartphones and other smaller electronics. A basic business model is to provide a seller an instant online quote based on laptop characteristics, then to send a shipping label and prepaid box to the seller, to erase, reformat, and process the laptop, and to pay rapidly by cheque. A majority of these companies are also generalized electronic waste recyclers as well; organizations that recycle computers exclusively include Cash For Laptops, a laptop refurbisher in Nevada that claims to be the first to buy laptops online, in 2001.

Donations/Nonprofits

With the constant rising costs due to inflation, many families or schools do not have the sufficient funds available for computers to be utilized along with education standards. Families also impacted by disaster suffer as well due to the financial impact of the situation they have incurred. Many nonprofit organizations, such as InterConnection.org, can be found locally as well as around the web and give detailed descriptions as to what methods are used for dissemination and detailed instructions on how to donate. The impact can be seen locally and globally, affecting thousands of those in need. In Canada non profit organizations engaged in computer recycling, such as The Electronic

Recycling Association Calgary, Edmonton, Vancouver, Winnipeg, Toronto, Montreal, Computers for Schools Canada wide, are very active in collecting and refurbishing computers and laptops to help the non profit and charitable sectors and schools.

History

Although consumer electronics such as the radio have been popular since the 1920s, recycling was almost unheard of until the early 1990s. At the end of the 1970s the accelerating pace of domestic consumer electronics drastically shortened the lifespan of electronics such as TVs, VCRs and audio. New innovations appeared more quickly, making older equipment considered obsolete. Increased complexity and sophistication of manufacture made local repair more difficult. The retail market shifted gradually, but substantially from a few high-value items that were cherished for years and repaired when necessary, to short-lived items that were rapidly replaced owing to wear or simply fashion, and discarded rather than repaired. This was particularly evident in computing, highlighted by Moore's Law. In 1988 two severe incidents highlighted the approaching e-waste crisis. The cargo barge Khian Sea, was loaded with more than 14,000 tons of toxic ash from Pennsylvania which had been refused acceptance in New Jersey and the Caribbean. After sailing for 16 months, all the waste was dumped as "topsoil fertiliser" in Haiti and in the Bay of Bengal by November 1988. In June 1988, a large illegal toxic waste dump which had been created by an Italian company was discovered. This led to the formation of the Basel Convention to stem the flow of poisonous substances from developed countries in 1989.

In 1991, the first electronic waste recycling system was implemented in Switzerland, beginning with collection of old refrigerators but gradually expanding to cover all devices. The organisation SWICO handles the programme, and is a partnership between IT retailers.

The first publication to report the recycling of computers and electronic waste was published on the front page of the New York Times on April 14, 1993 by columnist Steve Lohr. It detailed the work of Advanced Recovery Inc., a small recycler, in trying to safely dismantle computers, even if most waste was landfilled. Several other companies emerged in the early 1990s, chiefly in Europe, where national 'take back' laws compelled retailers to use them.

After these schemes were set up, many countries did not have the capacity to deal with the sheer quantity of e-waste they generated or its hazardous nature. They began to export the problem to developing countries without enforced environmental legislation. This is cheaper: the cost of recycling of computer monitors in the US is ten times more than in China. Demand in Asia for electronic waste began to grow when scrap yards found they could extract valuable substances such as copper, iron, silicon, nickel and gold, during the recycling process.

The Waste Electrical and Electronic Equipment Directive (WEEE Directive) became European Law in February 2003 and covers all aspects of recycling all types of appliance. This was followed by Electronic Waste Recycling Act, enshrined in Californian law in January 2005

The 2000s saw a large increase in both the sale of electronic devices and their growth as a waste stream: in 2002 e-waste grew faster than any other type of waste in the EU. This caused investment in modern, automated facilities to cope with the influx of redundant appliances.

E-cycling

"E-cycling" or "E-waste" is an initiative by the United States Environmental Protection Agency (EPA) which refers to donations, reuse, shredding and general collection of used electronics. Generically, the term refers to the process of collecting, brokering, disassembling, repairing and recycling the components or metals contained in used or discarded electronic equipment, otherwise known as electronic waste (e-waste). "E-cyclable" items include, but are not limited to: televisions, computers, microwave ovens, vacuum cleaners, telephones and cellular phones, stereos, and VCRs and DVDs just about anything that has a cord, light or takes some kind of battery.

Investment in e-cycling facilities has been increasing recently due to technology's rapid rate of obsolescence, concern over improper methods, and opportunities for manufacturers to influence the secondary market (used and reused products). The higher metal prices is also having more recycling taking place. The controversy around methods stems from a lack of agreement over preferred outcomes.

World markets with lower disposable incomes, consider 75% repair and reuse to be valuable enough to justify 25% disposal. Debate and certification standards may be leading to better definitions, though civil law contracts, governing the expected process are still vital to any contracted process, as poorly defined as "e-cycling".

Pros of E-cycling

The e-waste disposal occurring after processing for reuse, repair of equipment, and recovery of metals may be unethical or illegal when e-scrap of many kinds is transported overseas to developing countries for such processing. It is transported as if to be repaired and/or recycled, but after processing the less valuable e-scrap becomes e-waste/pollution there. Another point of view is that the net environmental cost must be compared to and include the mining, refining and extraction with its waste and pollution cost of new products manufactured to replace secondary products which are routinely destroyed in wealthier nations, and which cannot economically be repaired in older or obsolete products. As an example of negative impacts of e-waste, pollution of groundwater has become so serious in areas surrounding China's landfills that water must be shipped in from 18 miles (29 km) away. However, mining of new metals can have even broader impacts on groundwater. Either thorough e-cycling processing, domestic processing or overseas repair, can help the environment by avoiding pollution. Such e-cycling can theoretically be a sustainable alternative to disposing of e-waste in landfills. In addition, e-cycling allows for the reclamation of potential conflict minerals, like gold and wolframite, which requires less of those to be mined and lessens the potential money flow to militias and other exploitative actors in third-world that profit from mining them.

Supporters of one form of "required e-cycling" legislation argue that e-cycling saves taxpayers money, as the financial responsibility would be shifted from the taxpayer to the manufacturers. Advocates of more simple legislation (such as landfill bans for e-waste) argue that involving manufacturers does not reduce the cost to consumers, if reuse value is lost, and the resulting costs are then passed on to consumers in new products, particularly affecting markets which can hardly afford new products. It is theorized that manufacturers who take part in e-cycling would be motivated to use fewer materials in the production process, create longer lasting products, and implement safer, more efficient recycling systems. This theory is sharply disputed and has never been demonstrated.

Criticisms of E-cycling

The critics of e-cycling are just as vocal as its advocates. According to the Reason Foundation, e-cycling only raises the product and waste management costs of e-waste for consumers and limits innovation on the part of high-tech companies. They also believe that e-cycling facilities could unintentionally cause great harm to the environment. Critics claim that e-waste doesn't occupy a significant portion of total waste. According to a European study, only 4% of waste is electronic.

Another opposition to e-cycling is that many problems are posed in disassembly: the process is costly and dangerous because of the heavy metals of which the electronic products are composed, and as little as 1-5% of the original cost of materials can be retrieved. A final problem that people find is that identity fraud is all too common in regards to the disposal of electronic products. As the programs are legislated, creating winners and losers among e-cyclers with different locations and processes, it may be difficult to distinguish between criticism of e-cycling as a practice, and criticism of the specific legislated means proposed to enhance it.

The Fate of E-waste

A hefty criticism often lobbed at reuse based recyclers is that people think that they are recycling their electronic waste, when in reality it is actually being exported to developing countries like China, India, and Nigeria. For instance, at free recycling drives, "recyclers" may not be staying true to their word, but selling e-waste overseas or to parts brokers. Studies indicate that 50-80% of the 300,000 to 400,000 tons (270,000 to 360,000 tonnes) of e-waste is being sent overseas, and that approximately 2 million tons (1.8 million tonnes) per year go to U.S. landfills.

Although not possible in all circumstances, the best way to e-cycle is to upcycle e-waste. On the other hand, the electronic products in question are generally manufactured, and repaired under warranty, in the same nations, which anti-reuse recyclers depict as primitive. Reuse-based e-recyclers believe that fair-trade incentives for export markets will lead to better results than domestic shredding. There has been a continued debate between export-friendly e-cycling and increased regulation of that practice.

In the European Union, debate regarding the export of e-waste has resulted in a significant amendment to the WEEE directive (January 2012) with a view to significantly diminishing the export of WEEE (untreated e-waste). During debate in Strasburg, MEPs stated that "53 million tonnes of WEEE were generated in 2009 but only 18% collected for recycling" with the remainder being exported or sent to landfill. The Amendment, voted through by a unanimous 95% of representatives, removed the re-use (repair and refurbishmet) aspect of the directive and placed more emphasis upon recycling and recovery of precious metals and base metals. The changes went further by placing the burden upon registered exporters to prove that used equipment leaving Europe was "fit for purpose".

Policy Issues and Current Efforts

Currently, pieces of government legislation and a number of grassroots efforts have contributed to the growth of e-cycling processes which emphasize decreased exports over increased reuse rates. The Electronic Waste Recycling Act was passed in California in 2003. It requires that consumers pay an extra fee for certain types of electronics, and the collected money be then redistributed to recycling companies that are qualified to properly recycle these products. It is the only state that legislates against e-waste through this kind of consumer fee; the other states' efforts focus on producer responsibility laws or waste disposal bans. No study has shown that per capita recovery is greater in one type of legislated program (e.g. California) versus ordinary waste disposal bans (e.g. Massachusetts), though recovery has greatly increased in states which use either method.

As of September, 2006, Dell developed the nation's first completely free recycling program, furthering the responsibilities that manufacturers are taking for e-cycling. Manufacturers and retailers such as Best Buy, Sony, and Samsung have also set up recycling programs. This program does not accept televisions, which are the most expensive used electronic item, and are unpopular in markets which must deal with televisions when the more valuable computers have been cherry picked.

Another step being taken is the recyclers' pledge of true stewardship, sponsored by the Computer TakeBack Campaign. It has been signed by numerous recyclers promising to recycle responsibly. Grassroots efforts have also played a big part in this issue, as they and other community organizations are being formed to help responsibly recycle e-waste. Other grassroots campaigns are Basel, the Computer TakeBack Campaign (co-coordinated by the Grassroots Recycling Network), and the Silicon Valley Toxics Coalition. No study has shown any difference in recycling methods under the Pledge, and no data is available to demonstrate difference in management between "Pledge" and non-Pledge companies, though it is assumed that the risk of making false claims will prevent Pledge companies from wrongly describing their processes.

Many people believe that the U.S. should follow the European Union model in regards to its management of e-waste. In this program, a directive forces manufacturers to take responsibility for e-cycling; it also demands manufacturers' mandatory take-back and places bans on exporting e-waste to developing countries. Another longer-term solution is for computers to be composed of less dangerous products and many people disagree. No data has been provided to show that people who agree with the European model have based their agreement on measured outcomes or experience-based scientific method.

Data Security

Electronic waste dump at Agbogbloshie, Ghana. Organized criminals commonly search the drives for information to use in local scams.

E-waste presents a potential security threat to individuals and exporting countries. Hard drives that are not properly erased before the computer is disposed of can be reopened, exposing sensitive information. Credit card numbers, private financial data, account information and records of online transactions can be accessed by most willing individuals. Organized criminals in Ghana commonly search the drives for information to use in local scams.

Government contracts have been discovered on hard drives found in Agbogbloshie, Ghana. Multimillion-dollar agreements from United States security institutions such as the Defense Intelligence Agency (DIA), the Transportation Security Administration and Homeland Security have all resurfaced in Agbogbloshie.

Reasons to Destroy and Recycle Securely

There are ways to ensure that not only hardware is destroyed but also the private data on the hard drive. Having customer data stolen, lost, or misplaced contributes to the ever growing number of people who are affected by identity theft, which can cause corporations to lose more than just money. The image of a company that holds secure data, such as banks, law firms, pharmaceuticals, and credit corporations is also at risk. If a company's public image is hurt, it could cause consumers to not use their services and could cost millions in business losses and positive public relation campaigns. The cost of data breaches "varies widely, ranging from $90 to $50,000 (under HIPAA's new HITECH amendment, that came about through the American Recovery and Revitalization act of 2009), as per customer record, depending on whether the breach is "low-profile" or "high-profile" and the company is in a non-regulated or highly regulated area, such as banking or medical institutions."

There is also a major backlash from the consumer if there is a data breach in a company that is supposed to be trusted to protect their private information. If an organization has any consumer info on file, they must by law (Red Flags Clarification act of 2010) have written information protection policies and procedures in place, that serve to combat, mitigate, and detect vulnerable areas that could result in identity theft. The United States Department of Defense has published a standard to which recyclers and individuals may meet in order to satisfy HIPAA requirements.

Secure Recycling

Countries have developed standards, aimed at businesses and with the purpose of ensuring the security of Data contained in 'confidential' computer media [NIST 800-88: US standard for Data Remenance][HMG CESG IS5, Baseline & Enhanced, UK Government Protocol for Data Destruction]. National Association for Information Destruction (NAID) "is the international trade association for companies providing information destruction services. Suppliers of products, equipment and services to destruction companies are also eligible for membership. NAID's mission is to promote the information destruction industry and the standards and ethics of its member companies." There are companies that follow the guidelines from NAID and also meet all Federal EPA and local DEP regulations.

The typical process for computer recycling aims to securely destroy hard drives while still recycling the byproduct. A typical process for effective computer recycling:

1. Receive hardware for destruction in locked and securely transported vehicles.

2. Shred hard drives.

3. Separate all aluminum from the waste metals with an electromagnet.

4. Collect and securely deliver the shredded remains to an aluminum recycling plant.

5. Mold the remaining hard drive parts into aluminum ingots.

The Asset Disposal and Information Security Alliance (ADISA) publishes an *ADISA IT Asset Disposal Security Standard* that covers all phases of the e-waste disposal process from collection to transportation, storage and sanitization's at the disposal facility. It also conducts periodic audits of disposal vendors.

Computer Recycling Techniques

Data Shredder

Data Shredder (also known as CBL Data Shredder) is a data destruction utility designed to securely erase a hard disk or digital storage device, completely removing the data and making it unrecoverable. The software utilizes an overwrite method of destroying data rather than other means of data destruction (such as: degaussing, physical destruction).

Wiping Standards

	Windows Version	DOS Version
United States Department of Defense 5220.22-M National Industrial Security Standard	Yes	Yes
Germany BSI Verschlusssachen-IT-Richtlinien (VSITR (de))	Yes	Yes
Bruce Schneier's Algorithm	Yes	Yes
Peter Gutmann's Algorithm	Yes	Yes
RCMP DSX Method	Yes	Yes
Custom Method (zeroes, ones, random, number of passes)	Yes	Yes

International Data Destruction and Demand

Widespread need for secure data destruction meant that ltanguage-variants of the program needed to be developed to support demand. Coverage of the product in countries like Germany and Singapore lead to excessive downloading and general popularity. PC Magazine included Data Shredder as one of its must-have security tools. A large increase in electronic waste due to high quanitites of old computers being thrown out as garbage in dumps and landfills made security of data stored on hard drives a bigger concern of the general public. Data destruction became a part of regular service offerings from the company as press coverage highlighted the step in the data recovery process.

Data Erasure

Data erasure (also called data clearing or data wiping) is a software-based method of overwriting the data that aims to completely destroy all electronic data residing on a hard disk drive or other digital media. Permanent data erasure goes beyond basic file deletion commands, which only remove direct pointers to the data disk sectors and make the data recovery possible with common software tools. Unlike degaussing and physical destruction, which render the storage media unusable, data erasure removes all information while leaving the disk operable, preserving IT assets and the environment. New flash memory–based media implementations, such as solid-state drives or USB flash drives can cause data erasure techniques to fail allowing remnant data to be recoverable.

Software-based overwriting uses a software application to write a stream of zeros, ones or meaningless pseudorandom data onto all sectors of a hard disk drive. There are key differentiators between data erasure and other overwriting methods, which can leave data intact and raise the risk of data breach, identity theft or failure to achieve regulatory compliance. Many data eradication programs also provide multiple overwrites so that they support recognized government and industry standards, though a single-pass overwrite is widely considered to be sufficient for modern hard disk drives. Good software should provide verification of data removal, which is necessary for meeting certain standards.

To protect the data on lost or stolen media, some data erasure applications remotely destroy the data if the password is incorrectly entered. Data erasure tools can also target specific data on a disk for routine erasure, providing a hacking protection method that is less time-consuming than software encryption. Hardware/firmware encryption built into the drive itself or integrated controllers is a popular solution with no degradation in performance at all.

Presently, dedicated hardware/firmware encryption solutions can perform a 256-bit full AES encryption faster than the drive electronics can write the data. Drives with this capability are known as self-encrypting drives (SEDs); they are present on most modern enterprise-level laptops and are increasingly used in the enterprise to protect the data. Changing the encryption key renders inaccessible all data stored on a SED, which is an easy and very fast method for achieving a 100% data erasure. Theft of an SED results in a physical asset loss, but the stored data is inaccessible without the decryption key that is not stored on a SED, assuming there are no effective attacks against AES or its implementation in the drive hardware.

Importance

Information technology (IT) assets commonly hold large volumes of confidential data. Social security numbers, credit card numbers, bank details, medical history and classified information are often stored on computer hard drives or servers. These can inadvertently or intentionally make their way onto other media such as printer, USB, flash, Zip, Jaz, and REV drives.

Data Breach

Increased storage of sensitive data, combined with rapid technological change and the shorter lifespan of IT assets, has driven the need for permanent data erasure of electronic devices as they are retired or refurbished. Also, compromised networks and laptop theft and loss, as well as that of other portable media, are increasingly common sources of data breaches.

If data erasure does not occur when a disk is retired or lost, an organization or user faces a possibility that the data will be stolen and compromised, leading to identity theft, loss of corporate reputation, threats to regulatory compliance and financial impacts. Companies have spent nearly $5 million on average to recover when corporate data were lost or stolen.High profile incidents of data theft include:

- CardSystems Solutions (2005-06-19): Credit card breach exposes 40 million accounts.

- Lifeblood (2008-02-13): Missing laptops contain personal information including dates of birth and some Social Security numbers of 321,000.

- Hannaford (2008-03-17): Breach exposes 4.2 million credit, debit cards.

- Compass Bank (2008-03-21): Stolen hard drive contains 1,000,000 customer records.

- University of Florida College of Medicine, Jacksonville (2008-05-20): Photographs and identifying information of 1,900 on improperly disposed computer.

- Oklahoma Corporation Commission (2008-05-21): Server sold at auction compromises more than 5,000 Social Security numbers.

Regulatory Compliance

Strict industry standards and government regulations are in place that force organizations to mitigate the risk of unauthorized exposure of confidential corporate and government data. Regulations in the United States include HIPAA (Health Insurance Portability and Accountability Act); FACTA (The Fair and Accurate Credit Transactions Act of 2003); GLB (Gramm-Leach Bliley); Sarbanes-Oxley Act (SOx); and Payment Card Industry Data Security Standards (PCI DSS) and the Data Protection Act in the United Kingdom. Failure to comply can result in fines and damage to company reputation, as well as civil and criminal liability.

Preserving Assets and the Environment

Data erasure offers an alternative to physical destruction and degaussing for secure removal of all the disk data. Physical destruction and degaussing destroy the digital media, requiring disposal and contributing to electronic waste while negatively impacting the carbon footprint of individuals and companies. Hard drives are nearly 100% recyclable and can be collected at no charge from a variety of hard drive recyclers after they have been sanitized.

Limitations

Data erasure may not work completely on flash based media, such as Solid State Drives and USB Flash Drives, as these devices can store remnant data which is inaccessible to the erasure technique, and data can be retrieved from the individual flash memory chips inside the device. Data erasure through overwriting only works on hard drives that are functioning and writing to all sectors. Bad sectors cannot usually be overwritten, but may contain recoverable information. Bad sectors, however, may be invisible to the host system and thus to the erasing software. Disk encryption before use prevents this problem. Software-driven data erasure could also be compromised by malicious code.

Differentiators

Software-based data erasure uses a special application to write a combination of ones and zeroes onto each hard disk drive sector. The level of security depends on the number of times the entire drive is written over.

Full Disk Overwriting

While there are many overwriting programs, only those capable of complete data erasure offer full security by destroying the data on all areas of a hard drive. Disk overwriting programs that cannot access the entire hard drive, including hidden/locked areas like the host protected area (HPA), device configuration overlay (DCO), and remapped sectors, perform an incomplete erasure, leaving some of the data intact. By accessing the entire hard drive, data erasure eliminates the risk of data remanence.

Data erasure can also bypass the BIOS and OS. Overwriting programs that operate through the BIOS and OS will not always perform a complete erasure due to altered or corrupted BIOS data and may report back a complete and successful erasure even if they do not access the entire hard disk, leaving the data accessible.

Hardware Support

Data erasure can be deployed over a network to target multiple PCs rather than having to erase each one sequentially. In contrast with DOS-based overwriting programs that may not detect all network hardware, Linux-based data erasure software supports high-end server and storage area network (SAN) environments with hardware support for Serial ATA, Serial Attached SCSI (SAS) and Fibre Channel disks and remapped sectors. It operates directly with sector sizes such as 520, 524, and 528, removing the need to first reformat back to 512 sector size.

Standards

Many government and industry standards exist for software-based overwriting that removes the data. A key factor in meeting these standards is the number of times the data are overwritten. Also, some standards require a method to verify that all the data have been removed from the entire hard drive and to view the overwrite pattern. Complete data erasure should account for hidden areas, typically DCO, HPA and remapped sectors.

The 1995 edition of the National Industrial Security Program Operating Manual (DoD 5220.22-M) permitted the use of overwriting techniques to sanitize some types of media by writing all addressable locations with a character, its complement, and then a random character. This provision was removed in a 2001 change to the manual and was never permitted for Top Secret media, but it is still listed as a technique by many providers of the data erasure software.

Data erasure software should provide the user with a validation certificate indicating that the overwriting procedure was completed properly. Data erasure software should also comply with requirements to erase hidden areas, provide a defects log list and list bad sectors that could not be overwritten.

Overwriting Standard	Date	Overwriting Rounds	Pattern	Notes
U.S. Navy Staff Office Publication NAVSO P-5239-26	1993	3	A character, its complement, random	Verification is mandatory
U.S. Air Force System Security Instruction 5020	1996	3	All zeros, all ones, any character	Verification is mandatory
Peter Gutmann's Algorithm	1996	1 to 35	Various, including all of the other listed methods	Originally intended for MFM and RLL disks, which are now obsolete
Bruce Schneier's Algorithm	1996	7	All ones, all zeros, pseudo-random sequence five times	
U.S. DoD Unclassified Computer Hard Drive Disposition	2001	3	A character, its complement, another pattern	
German Federal Office for Information Security	2004	2-3	Non-uniform pattern, its complement	
Communications Security Establishment Canada ITSG-06	2006	3	All ones or zeros, its complement, a pseudo-random pattern	For unclassified media
NIST SP-800-88	2006	1	?	

U.S. National Industrial Security Program Operating Manual (DoD 5220.22-M)	2006	?	?	No longer specifies any method.
NSA/CSS Storage Device Declassification Manual (SDDM)	2007	0	?	Degauss or destroy only
Australian Government ICT Security Manual 2014 - Controls	2014	1	Random pattern (only for disks bigger than 15 GB)	Degauss magnetic media or destroy Top Secret media
New Zealand Government Communications Security Bureau NZSIT 402	2008	1	?	For data up to Confidential
British HMG Infosec Standard 5, Baseline Standard	?	1	Random Pattern	Verification is mandatory
British HMG Infosec Standard 5, Enhanced Standard	?	3	All ones, all zeros, random	Verification is mandatory

Data can sometimes be recovered from a broken hard drive. However, if the platters on a hard drive are damaged, such as by drilling a hole through the drive (and the platters inside), then the data can only theoretically be recovered by bit-by-bit analysis of each platter with advanced forensic technology.

Number of Overwrites Needed

Data on floppy disks can sometimes be recovered by forensic analysis even after the disks have been overwritten once with zeros (or random zeros and ones). This is not the case with modern hard drives:

- According to the 2006 NIST Special Publication 800-88 Section 2.3 (p. 6): "Basically the change in track density and the related changes in the storage medium have created a situation where the acts of clearing and purging the media have converged. That is, for ATA disk drives manufactured after 2001 (over 15 GB) clearing by overwriting the media once is adequate to protect the media from both keyboard and laboratory attack."

- According to the 2006 Center for Magnetic Recording Research Tutorial on Disk Drive Data Sanitization Document (p. 8): "Secure erase does a single on-track erasure of the data on the disk drive. The U.S. National Security Agency published an Information Assurance Approval of single-pass overwrite, after technical testing at CMRR showed that multiple on-track overwrite passes gave no additional erasure." "Secure erase" is a utility built into modern ATA hard drives that overwrites all data on a disk, including remapped (error) sectors.

- Further analysis by Wright et al. seems to also indicate that one overwrite is all that is generally required.

E-waste and Information Security

E-waste presents a potential security threat to individuals and exporting countries. Hard drives that are not properly erased before the computer is disposed of can be reopened, exposing sensitive information. Credit card numbers, private financial data, account information and records of online transactions can be accessed by most willing individuals. Organized criminals in Ghana commonly search the drives for information to use in local scams.

Government contracts have been discovered on hard drives found in Agbogbloshie, the unregulated e-waste centre in Ghana. Multimillion-dollar agreements from United States security institutions such as the Defense Intelligence Agency (DIA), the Transportation Security Administration and Homeland Security have all resurfaced in Agbogbloshie.

Degaussing

Degaussing is the process of decreasing or eliminating a remnant magnetic field. It is named after the gauss, a unit of magnetism, which in turn was named after Carl Friedrich Gauss. Due to magnetic hysteresis, it is generally not possible to reduce a magnetic field completely to zero, so degaussing typically induces a very small "known" field referred to as bias. Degaussing was originally applied to reduce ships' magnetic signatures during the Second World War. Degaussing is also used to reduce magnetic fields in CRT monitors and to destroy data held on magnetic data storage.

Ships' Hulls

RMS *Queen Mary* arriving in New York Harbor, 20 June 1945, with thousands of U.S. soldiers – note the prominent degaussing coil running around the outer hull

The term was first used by then Cmdr Charles F. Goodeve, RCNVR, during World War II, while trying to counter the German magnetic mines that were playing havoc with the British fleet. The mines detected the increase in magnetic field when the steel in a ship concentrated the Earth's magnetic field over it. Admiralty scientists, including Goodeve, developed a number of systems to induce a small "N-pole up" field into the ship to offset this effect, meaning that the net field was the same as background. Since the Germans used the gauss as the unit of the strength of the magnetic field in their mines' triggers (this was not yet a standard measure), Goodeve referred to the various processes to counter the mines as "degaussing". The term became a common word.

The original method of degaussing was to install electromagnetic coils into the ships, known simply as coiling. In addition to being able to bias the ship continually, coiling also allowed the bias field to be reversed in the southern hemisphere, where the mines were set to detect "S-pole down" fields. British ships, notably cruisers and battleships, were well protected by about 1943.

Control panel of the *MES*-device (*"Magnetischer Eigenschutz"* German: magnetic self-protection) in a German submarine

Installing such special equipment was, however, far too expensive and difficult to service all ships that would need it, so the navy developed an alternative called wiping, which Goodeve also devised, and which is now also called deperming. This procedure simply dragged a large electrical cable along the side of the ship with a pulse of about 2000 amperes flowing through it. This induced the proper field into the ship in the form of a slight bias. It was originally thought that the pounding of the sea and the ship's engines would slowly randomize this field, but in testing, this was found not to be a real problem. A more serious problem was later realized: as a ship travels through Earth's magnetic field, it will slowly pick up that field, counteracting the effects of the degaussing. From then on captains were instructed to change direction as often as possible to avoid this problem. Nevertheless, the bias did wear off eventually, and ships had to be degaussed on a schedule. Smaller ships continued to use wiping through the war.

After the war, the capabilities of the magnetic fuses were greatly improved, by detecting not the field itself, but *changes* in it. This meant a degaussed ship with a magnetic "hot spot" would still set off the mine. Additionally, the precise orientation of the field was also measured, something a simple bias field could not remove, at least for all points on the ship. A series of ever-increasingly complex coils were introduced to offset these effects, with modern systems including no fewer than three separate sets of coils to reduce the field in all axes.

The US Navy tested, in April 2009, a prototype of its High-Temperature Superconducting Degaussing Coil System, referred to as "HTS Degaussing". The system works by encircling the vessel with superconducting ceramic cables whose purpose is to neutralize the ship's magnetic signature, as in the legacy copper systems. The main advantage of the HTS Degaussing Coil system is greatly reduced weight (sometimes by as much as 80%) and increased efficiency.

A sea-going metal-hulled ship or submarine, by its very nature, develops a magnetic signature as it travels, due to a magneto-mechanical interaction with Earth's magnetic field. It also picks up the magnetic orientation of the earth's magnetic field where it is built. This signature can be exploited by magnetic mines or facilitate the detection of a submarine by ships or aircraft with magnetic anomaly detection (MAD) equipment. Navies use the deperming procedure, in conjunction with degaussing, as a countermeasure against this.

Specialized deperming facilities, such as the United States Navy's Lambert's Point Deperming Station at Naval Station Norfolk, or Pacific Fleet Submarine Drive-In Magnetic Silencing Facility (MSF) at Joint Base Pearl Harbor-Hickam, are used to perform the procedure. During a closed-wrap magnetic treatment, heavy-gauge copper cables encircle the hull and superstructure of the vessel, and high electrical currents (up to 4000 amperes) are pulsed through the cables. This has the effect of "resetting" the ship's magnetic signature to the ambient level after flashing its hull with electricity. It is also possible to assign a specific signature that is best suited to the particular area of the world in which the ship will operate. In drive-in magnetic silencing facilities, all cables are either hung above, below and on the sides, or concealed within the structural elements of facilities. Deperming is "permanent". It is only done once unless major repairs or structural modifications are done to the ship.

During World War II, the United States Navy commissioned a specialized class of degaussing ships that were capable of performing this function. One of them, USS *Deperm* (ADG-10), was named after the procedure.

Early Experiments

With the introduction of iron ships, the negative effect of the metal hull on steering compasses was noted. It was also observed that lightning strikes had a significant effect on compass deviation, identified in some extreme cases as being caused by the reversal of the ship's magnetic signature. In 1866, Evan Hopkins of London registered a patent for a process "to depolarise iron vessels and leave them thenceforth free from any compass-disturbing influence whatever". The technique was described as follows: "For this purpose he employed a number of Grove's batteries and electromagnets. The latter were to be passed along the plates till the desired end had been obtained... the process must not be overdone for fear of re-polarising in the opposite direction." The invention was, however, reported to be "incapable of being carried to a successful issue", and "quickly died a natural death".

Monitors

Until recently, the most common use of degaussing was in CRT-based TV sets and computer monitors. For example, many monitors use a metal plate near the front of the tube to guide the electron beams from the back. This plate, the shadow mask, can pick up strong external fields and from that point produce discoloration on the display.

To minimize this, CRTs have a copper, or often in the case of cheaper appliances, aluminum, coil wrapped around the front of the display, known as the degaussing coil. Tubes without an internal coil can be degaussed using an external handheld version. Internal degaussing coils in CRTs are generally much weaker than external degaussing coils, since a better degaussing coil takes up more space. A degauss causes a magnetic field inside the tube to oscillate rapidly, with decreasing amplitude. This leaves the shadow mask with a small and somewhat randomized field, removing the discoloration.

Many televisions and monitors automatically degauss their picture tube when switched on, before an image is displayed. The high current surge that takes place during this automatic degauss is the cause of an audible "thunk" or loud hum, which can be heard (and felt) when televisions and CRT computer monitors are switched on. Visually, this causes the image to shake dramatically for a short period of time. A degauss option is also usually available for manual selection in the operations menu in such appliances.

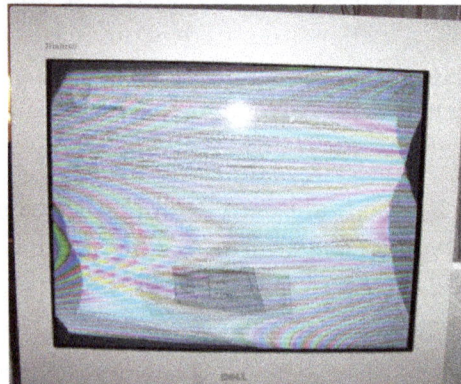

A degaussing in progress

In most commercial equipment the current surge to the degaussing coil is regulated by a simple PTC thermistor device, which initially has a low resistance but quickly changes to a high resistance due to the heating effect of the current flow. Such devices are designed for a one-off transition from cold to hot at power up, "experimenting" with the degauss effect by repeatedly switching the device on and off, may cause this component to fail. The effect will also be weaker, since the PTC will not have had time to cool off.

Magnetic Data Storage Media

Data is stored in the magnetic media, such as hard drives, floppy disks, and magnetic tape, by making very small areas called magnetic domains change their magnetic alignment to be in the direction of an applied magnetic field. This phenomenon occurs in much the same way a compass needle points in the direction of the Earth's magnetic field. Degaussing, commonly called erasure, leaves the domains in random patterns with no preference to orientation, thereby rendering previous data unrecoverable. There are some domains whose magnetic alignment is not randomized after degaussing. The information these domains represent is commonly called magnetic remanence or remanent magnetization. Proper degaussing will ensure there is insufficient magnetic remanence to reconstruct the data.

Erasure *via* degaussing may be accomplished in two ways: in AC erasure, the medium is degaussed by applying an alternating field that is reduced in amplitude over time from an initial high value (i.e., AC powered); in DC erasure, the medium is saturated by applying a unidirectional field (i.e., DC powered or by employing a permanent magnet). A degausser is a device that can generate a magnetic field for degaussing magnetic storage media.

Irreversible Damage to Some Media Types

Many forms of generic magnetic storage media can be reused after degaussing, including audio reel-to-reel tape, VHS videocassettes, and floppy disks. These older media types are simply a raw medium which are overwritten with fresh new patterns, created by fixed-alignment read/write heads.

For certain forms of computer data storage, however, such as modern hard drives and some tape backup drives, degaussing renders the magnetic media completely unusable and damages the storage system. This is due to the devices having an infinitely variable read/write head positioning mechanism which relies on special servo control data (e.g. Gray Code) that is meant to be permanently embedded into the magnetic media. This servo data is written onto the media a single time at the factory using special-purpose servo writing hardware.

The servo patterns are normally never overwritten by the device for any reason and are used to precisely position the read/write heads over data tracks on the media, to compensate for sudden jarring device movements, thermal expansion, or changes in orientation. Degaussing indiscriminately removes not only the stored data but also the servo control data, and without the servo data the device is no longer able to determine where data is to be read or written on the magnetic medium. The medium must be low-level formatted to become usable again; with modern hard drives, this is generally not possible without manufacturer-specific and often model-specific service equipment.

Data Remanence

Data remanence is the residual representation of digital data that remains even after attempts have been made to remove or erase the data. This residue may result from data being left intact by a nominal file deletion operation, by reformatting of storage media that does not remove data previously written to the media, or through physical properties of the storage media that allow previously written data to be recovered. Data remanence may make inadvertent disclosure of sensitive information possible should the storage media be released into an uncontrolled environment (*e.g.*, thrown in the trash, or lost).

Various techniques have been developed to counter data remanence. These techniques are classified as clearing, purging/sanitizing or destruction. Specific methods include overwriting, degaussing, encryption, and media destruction.

Effective application of countermeasures can be complicated by several factors, including media that are inaccessible, media that cannot effectively be erased, advanced storage systems that maintain histories of data throughout the data's life cycle, and persistence of data in memory that is typically considered volatile.

Several standards exist for the secure removal of data and the elimination of data remanence.

Causes

Many operating systems, file managers, and other software provide a facility where a file is not immediately deleted when the user requests that action. Instead, the file is moved to a holding area, to allow the user to easily revert a mistake. Similarly, many software products automatically create backup copies of files that are being edited, to allow the user to restore the original version, or to recover from a possible crash (*autosave* feature).

Even when an explicit deleted file retention facility is not provided or when the user does not use it, operating systems do not actually remove the contents of a file when it is deleted unless they are aware that explicit erasure commands are required, like on a solid-state drive. (In such cases, the operating system will issue the Serial ATA TRIM command or the SCSI UNMAP command to let the drive know to no longer maintain the deleted data.) Instead, they simply remove the file's entry from the file system directory, because this requires less work and is therefore faster, and the contents of the file—the actual data—remain on the storage medium. The data will remain there until the operating system reuses the space for new data. In some systems, enough filesystem metadata are also left behind to enable easy undeletion by commonly available utility software. Even when undelete has become impossible, the data, until it has been overwritten, can be read by software that reads disk sectors directly. Computer forensics often employs such software.

Likewise, reformatting, repartitioning, or reimaging a system is unlikely to write to every area of the disk, though all will cause the disk to appear empty or, in the case of reimaging, empty except for the files present in the image, to most software.

Finally, even when the storage media is overwritten, physical properties of the media may permit recovery of the previous contents. In most cases however, this recovery is not possible by just reading from the storage device in the usual way, but requires using laboratory techniques such as disassembling the device and directly accessing/reading from its components.

The section on complications gives further explanations for causes of data remanence.

Countermeasures

There are three levels commonly recognized for eliminating remnant data:

Clearing

Clearing is the removal of sensitive data from storage devices in such a way that there is assurance that the data may not be reconstructed using normal system functions or software file/data recovery utilities. The data may still be recoverable, but not without special laboratory techniques.

Clearing is typically an administrative protection against accidental disclosure within an organization. For example, before a hard drive is re-used within an organization, its contents may be cleared to prevent their accidental disclosure to the next user.

Purging

Purging or sanitizing is the removal of sensitive data from a system or storage device with the intent that the data can not be reconstructed by any known technique. Purging, proportional to the sensitivity of the data, is generally done before releasing media beyond control, such as before discarding old media, or moving media to a computer with different security requirements.

Destruction

The storage media is made unusable for conventional equipment. Effectiveness of destroying the media varies by medium and method. Depending on recording density of the media, and/or the destruction technique, this may leave data recoverable by laboratory methods. Conversely, destruction using appropriate techniques is the most secure method of preventing retrieval.

Specific Methods

Overwriting

A common method used to counter data remanence is to overwrite the storage media with new data. This is often called wiping or shredding a file or disk, by analogy to common methods of destroying print media, although the mechanism bears no similarity to these. Because such a method can often be implemented in software alone, and may be able to selectively target only part of the media, it is a popular, low-cost option for some applications. Overwriting is generally an acceptable method of clearing, as long as the media is writable and not damaged.

The simplest overwrite technique writes the same data everywhere—often just a pattern of all zeros. At a minimum, this will prevent the data from being retrieved simply by reading from the media again using standard system functions.

In an attempt to counter more advanced data recovery techniques, specific overwrite patterns and multiple passes have often been prescribed. These may be generic patterns intended to eradicate any trace signatures, for example, the seven-pass pattern: 0xF6, 0x00, 0xFF, random, 0x00, 0xFF, random; sometimes erroneously attributed to the US standard DOD 5220.22-M.

One challenge with an overwrite is that some areas of the disk may be inaccessible, due to media degradation or other errors. Software overwrite may also be problematic in high-security environments which require stronger controls on data commingling than can be provided by the software in use. The use of advanced storage technologies may also make file-based overwrite ineffective.

There are specialized machines and software that are capable of doing overwriting. The software can sometimes be a standalone operating system specifically designed for data destruction. There are also machines specifically designed to wipe hard drives to the department of defense specifications DOD 5220.22-M.

Feasibility of Recovering Overwritten Data

Peter Gutmann investigated data recovery from nominally overwritten media in the mid-1990s. He suggested magnetic force microscopy may be able to recover such data, and developed specific patterns, for specific drive technologies, designed to counter such. These patterns have come to be known as the Gutmann method.

Daniel Feenberg, an economist at the private National Bureau of Economic Research, claims that the chances of overwritten data being recovered from a modern hard drive amount to "urban legend". He also points to the "18½ minute gap" Rose Mary Woods created on a tape of Richard Nixon discussing the Watergate break-in. Erased information in the gap has not been recovered, and Feenberg claims doing so would be an easy task compared to recovery of a modern high density digital signal.

As of November 2007, the United States Department of Defense considers overwriting acceptable for clearing magnetic media within the same security area/zone, but not as a sanitization method. Only degaussing or physical destruction is acceptable for the latter.

On the other hand, according to the 2006 NIST Special Publication 800-88 (p. 7): "Studies have shown that most of today's media can be effectively cleared by one overwrite" and "for ATA disk drives manufactured after 2001 (over 15 GB) the terms clearing and purging have converged." An analysis by Wright et al. of recovery techniques, including magnetic force microscopy, also concludes that a single wipe is all that is required for modern drives. They point out that the long time required for multiple wipes "has created a situation where many organisations ignore the issue all together – resulting in data leaks and loss."

Degaussing

Degaussing is the removal or reduction of a magnetic field of a disk or drive, using a device called a degausser that has been designed for the media being erased. Applied to magnetic media, degaussing may purge an entire media element quickly and effectively.

Degaussing often renders hard disks inoperable, as it erases low-level formatting that is only done at the factory during manufacturing. In some cases, it is possible to return the drive to a functional state by having it serviced at the manufacturer. However, some modern degaussers use such a strong magnetic pulse that the motor that spins the platters may be destroyed in the degaussing process, and servicing may not be cost-effective. Degaussed computer tape such as DLT can generally be reformatted and reused with standard consumer hardware.

In some high-security environments, one may be required to use a degausser that has been approved for the task. For example, in US government and military jurisdictions, one may be required to use a degausser from the NSA's "Evaluated Products List".

Encryption

Encrypting data before it is stored on the media may mitigate concerns about data remanence. If the decryption key is strong and carefully controlled, it may effectively make any data on the media unrecoverable. Even if the key is stored on the media, it may prove easier or quicker to overwrite just the key, vs the entire disk. This process is called crypto erase in the security industry.

Encryption may be done on a file-by-file basis, or on the whole disk. Cold boot attacks are one of the few possible methods for subverting a full-disk encryption method, as there is no possibility of storing the plain text key in an unencrypted section of the medium.

Other side-channel attacks (such as keyloggers, acquisition of a written note containing the decryption key, or rubber hose cryptography) may offer a greater chance to success, but do not rely on weaknesses in the cryptographic method employed. As such, their relevance for this article is minor.

Media Destruction

Thorough destruction of the underlying storage media is the most certain way to counter data remanence. However, the process is generally time-consuming, cumbersome, and may require extremely thorough methods, as even a small fragment of the media may contain large amounts of data.

The pieces of a physically destroyed hard disk drive.

Specific destruction techniques include:

- Physically breaking the media apart (e.g., by grinding or shredding)

- Chemically altering the media into a non-readable, non-reverse-constructible state (e.g., through incineration or exposure to caustic/corrosive chemicals)

- Phase transition (e.g., liquefaction or vaporization of a solid disk)

- For magnetic media, raising its temperature above the Curie point

- For many electric/electronic volatile and non-volatile storage mediums, exposure to electromagnetic fields greatly exceeding safe operational specifications (e.g., high-voltage electric current or high-amplitude microwave radiation)

Complications

Inaccessible Media Areas

Storage media may have areas which become inaccessible by normal means. For example, magnetic disks may develop new bad sectors after data has been written, and tapes require inter-record gaps. Modern hard disks often feature reallocation of marginal sectors or tracks, automated in a way that the operating system would not need to work with it. The problem is especially significant in solid state drives (SSDs) that rely on relatively large relocated bad block tables. Attempts to counter data remanence by overwriting may not be successful in such situations, as data remnants may persist in such nominally inaccessible areas.

Advanced Storage Systems

Data storage systems with more sophisticated features may make overwrite ineffective, especially on a per-file basis. For example, journaling file systems increase the integrity of data by recording write operations in multiple locations, and applying transaction-like semantics; on such systems, data remnants may exist in locations "outside" the nominal file storage location. Some file systems also implement copy-on-write or built-in revision control, with the intent that writing to a file never overwrites data in-place. Furthermore, technologies such as RAID and anti-fragmentation techniques may result in file data being written to multiple locations, either by design (for fault tolerance), or as data remnants.

Wear leveling can also defeat data erasure, by relocating blocks between the time when they are originally written and the time when they are overwritten. For this reason, some security protocols tailored to operating systems or other software featuring automatic wear leveling recommend conducting a free-space wipe of a given drive and then copying many small, easily identifiable "junk" files or files containing other nonsensitive data to fill as much of that drive as possible, leaving only the amount of free space necessary for satisfactory operation of system hardware and software. As storage and/or system demands grow, the "junk data" files can be deleted as necessary to free up space; even if the deletion of "junk data" files is not secure, their initial nonsensitivity reduces to near zero the consequences of recovery of data remanent from them.

Optical Media

As optical media are not magnetic, they are not erased by conventional degaussing. Write-once optical media (CD-R, DVD-R, etc.) also cannot be purged by overwriting. Read/write optical media, such as CD-RW and DVD-RW, may be receptive to overwriting. Methods for successfully sanitizing optical discs include delaminating or abrading the metallic data layer, shredding, incinerating, destructive electrical arcing (as by exposure to microwave energy), and submersion in a polycarbonate solvent (e.g., acetone).

Data on Solid-state Drives

Research from the Center for Magnetic Recording and Research, University of California, San Diego has uncovered problems inherent in erasing data stored on solid-state drives (SSDs). Researchers discovered three problems with file storage on SSDs:

First, built-in commands are effective, but manufacturers sometimes implement them incorrectly. Second, overwriting the entire visible address space of an SSD twice is usually, but not always, sufficient to sanitize the drive. Third, none of the existing hard drive-oriented techniques for individual file sanitization are effective on SSDs.

Solid-state drives, which are flash-based, differ from hard-disk drives in two ways: first, in the way data is stored; and second, in the way the algorithms are used to manage and access that data. These differences can be exploited to recover previously erased data. SSDs maintain a layer of indirection between the logical addresses used by computer systems to access data and the internal addresses that identify physical storage. This layer of indirection hides idiosyncratic media interfaces and enhances SSD performance, reliability, and lifespan; but it can also produce copies of the data that are invisible to the user and that a sophisticated attacker could recover. For sanitizing entire disks, sanitize commands built into the SSD hardware have been found to be effective when implemented correctly, and software-only techniques for sanitizing entire disks have been found to work most, but not all, of the time. In testing, none of the software techniques were effective for sanitizing individual files. These included well-known algorithms such as the Gutmann method, US DoD 5220.22-M, RCMP TSSIT OPS-II, Schneier 7 Pass, and Mac OS X Secure Erase Trash.

The TRIM feature in many SSD devices, if properly implemented, will eventually erase data after it is deleted, but the process can take some time, typically several minutes. Many older operating systems do not support this feature, and not all combinations of drives and operating systems work.

Data in RAM

Data remanence has been observed in static random-access memory (SRAM), which is typically considered volatile (*i.e.*, the contents degrade with loss of external power). In one study, data retention was observed even at room temperature.

Data remanence has also been observed in dynamic random-access memory (DRAM). Modern DRAM chips have a built-in self-refresh module, as they not only require a power supply to retain data, but must also be periodically refreshed to prevent their data contents from fading away from the capacitors in their integrated circuits. A study found data remanence in DRAM with data retention of seconds to minutes at room temperature and "a full week without refresh when cooled with liquid nitrogen." The study authors were able to use a cold boot attack to recover cryptographic keys for several popular full disk encryption systems, including Microsoft BitLocker, Apple FileVault, dm-crypt for Linux, and TrueCrypt.

Despite some memory degradation, authors of the above described study were able to take advantage of redundancy in the way keys are stored after they have been expanded for efficient use, such as in key scheduling. The authors recommend that computers be powered down, rather than be left in a "sleep" state, when not in physical control of the owner. In some cases, such as certain modes of the software program BitLocker, the authors recommend that a boot password or a key on a removable USB device be used. TRESOR is a kernel patch for Linux specifically intended to prevent cold boot attacks on RAM by ensuring encryption keys are neither user accessible nor stored in RAM.

Electronic Waste

Electronic waste or e-waste describes discarded electrical or electronic devices. Used electronics which are destined for reuse, resale, salvage, recycling or disposal are also considered e-waste. Informal processing of e-waste in developing countries can lead to adverse human health effects and environmental pollution.

Defective and obsolete electronic equipment

Electronic scrap components, such as CPUs, contain potentially harmful components such as lead, cadmium, beryllium, or brominated flame retardants. Recycling and disposal of e-waste may involve significant risk to workers and communities in developed countries and great care must be taken to avoid unsafe exposure in recycling operations and leaking of materials such as heavy metals from landfills and incinerator ashes.

Definition

Hoarding (left), disassembling (center) and collecting (right) electronic waste in Bengaluru, India

"Electronic waste" may be defined as discarded computers, office electronic equipment, entertainment device electronics, mobile phones, television sets, and refrigerators. This includes used electronics which are destined for reuse, resale, salvage, recycling, or disposal. Others are re-usables (working and repairable electronics) and secondary scrap (copper, steel, plastic, etc.) to be "commodities", and reserve the term "waste" for residue or material which is dumped by the buyer rather than recycled, including residue from reuse and recycling operations. Because loads of surplus electronics are frequently commingled (good, recyclable, and non-recyclable), several public policy advocates apply the term "e-waste" broadly to all surplus electronics. Cathode ray tubes (CRTs) are considered one of the hardest types to recycle.

CRTs have relatively high concentration of lead and phosphors both of which are necessary for the display. The United States Environmental Protection Agency (EPA) includes discarded CRT monitors in its category of "hazardous household waste" but considers CRTs that have been set aside for testing to be commodities if they are not discarded, speculatively accumulated, or left unprotected from weather and other damage.

The EU and its member states operate a system via the European Waste Catalogue (EWC)- a European Council Directive, which is interpreted into "member state law". In the UK, this is in the form of the List of Wastes Directive. However, the list (and EWC) gives broad definition (EWC Code 16 02 13*) of Hazardous Electronic wastes, requiring "waste operators" to employ the Hazardous Waste Regulations (Annex 1A, Annex 1B) for refined definition. Constituent materials in the waste also require assessment via the combination of Annex II and Annex III, again allowing operators to further determine whether a waste is hazardous.

Debate continues over the distinction between "commodity" and "waste" electronics definitions. Some exporters are accused of deliberately leaving difficult-to-recycle, obsolete, or non-repairable equipment mixed in loads of working equipment (though this may also come through ignorance, or to avoid more costly treatment processes). Protectionists may broaden the definition of "waste" electronics in order to protect domestic markets from working secondary equipment.

The high value of the computer recycling subset of electronic waste (working and reusable laptops, desktops, and components like RAM) can help pay the cost of transportation for a larger number of worthless pieces than can be achieved with display devices, which have less (or negative) scrap value. In A 2011 report, "Ghana E-Waste Country Assessment", found that of 215,000 tons of electronics imported to Ghana, 30% were brand new and 70% were used. Of the used product, the study concluded that 15% was not reused and was scrapped or discarded. This contrasts with published but uncredited claims that 80% of the imports into Ghana were being burned in primitive conditions.

Amount of Electronic Waste World-wide

Rapid changes in technology, changes in media (tapes, software, MP3), falling prices, and planned obsolescence have resulted in a fast-growing surplus of electronic waste around the globe. Technical solutions are available, but in most cases a legal framework, a collection, logistics, and other services need to be implemented before a technical solution can be applied.

Display units (CRT, LCD, LED monitors), processors (CPU, GPU, or APU chips), memory (DRAM or SRAM), and audio components have different useful lives. Processors are most frequently out-dated (by software no longer being optimized) and are more likely to become "e-waste", while display units are most often replaced while working without repair attempts, due to changes in wealthy nation appetites for new display technology. This problem could potentially be solved with Modular Smartphones or Phonebloks. These types of phones are more durable and have the technology to change certain parts of the phone making them more environmentally friendly. Being able to simply replace the part of the phone that is broken will reduce e-waste. An estimated 50 million tons of E-waste are produced each year. The USA discards 30 million computers each year and 100 million phones are disposed of in Europe each year. The Environmental Protection Agency estimates that only 15–20% of e-waste is recycled, the rest of these electronics go directly into landfills and incinerators.

A fragment of discarded circuit board.

According to a report by UNEP titled, "Recycling – from E-Waste to Resources," the amount of e-waste being produced – including mobile phones and computers – could rise by as much as 500 percent over the next decade in some countries, such as India. The United States is the world leader in producing electronic waste, tossing away about 3 million tons each year. China already produces about 2.3 million tons (2010 estimate) domestically, second only to the United States. And, despite having banned e-waste imports, China remains a major e-waste dumping ground for developed countries.

Society today revolves around technology and by the constant need for the newest and most high tech products we are contributing to mass amount of e-waste. Since the invention of the iPhone, cell phones have become the top source of e-waste products because they are not made to last more than two years. Electrical waste contains hazardous but also valuable and scarce materials. Up to 60 elements can be found in complex electronics. As of 2013, Apple has sold over 796 million iDevices (iPod, iPhone, iPad). Cell phone companies make cell phones that are not made to last so that the consumer will purchase new phones. Companies give these products such short life spans because they know that the consumer will want a new product and will buy it if they make it. In the United States, an estimated 70% of heavy metals in landfills comes from discarded electronics.

While there is agreement that the number of discarded electronic devices is increasing, there is considerable disagreement about the relative risk (compared to automobile scrap, for example), and strong disagreement whether curtailing trade in used electronics will improve conditions, or make them worse. According to an article in *Motherboard*, attempts to restrict the trade have driven reputable companies out of the supply chain, with unintended consequences.

Global Trade Issues

One theory is that increased regulation of electronic waste and concern over the environmental harm in mature economies creates an economic disincentive to remove residues prior to export. Critics of trade in used electronics maintain that it is still too easy for brokers calling themselves recyclers to export unscreened electronic waste to developing countries, such as China, India and parts of Africa, thus avoiding the expense of removing items like bad cathode ray tubes (the processing of which is expensive and difficult). The developing countries have become toxic dump yards of e-waste. Proponents of international trade point to the success of fair trade programs in other industries, where cooperation has led to creation of sustainable jobs, and can bring affordable technology in countries where repair and reuse rates are higher.

Defenders of the trade in used electronics say that extraction of metals from virgin mining has been shifted to developing countries. Recycling of copper, silver, gold, and other materials from discarded electronic devices is considered better for the environment than mining. They also state that repair and reuse of computers and televisions has become a "lost art" in wealthier nations, and that refurbishing has traditionally been a path to development.

South Korea, Taiwan, and southern China all excelled in finding "retained value" in used goods, and in some cases have set up billion-dollar industries in refurbishing used ink cartridges, single-use cameras, and working CRTs. Refurbishing has traditionally been a threat to established manufacturing, and simple protectionism explains some criticism of the trade. Works like "The Waste Makers" by Vance Packard explain some of the criticism of exports of working product, for example the ban on import of tested working Pentium 4 laptops to China, or the bans on export of used surplus working electronics by Japan.

Opponents of surplus electronics exports argue that lower environmental and labor standards, cheap labor, and the relatively high value of recovered raw materials leads to a transfer of pollution-generating activities, such as smelting of copper wire. In China, Malaysia, India, Kenya, and various African countries, electronic waste is being sent to these countries for processing, sometimes illegally. Many surplus laptops are routed to developing nations as "dumping grounds for e-waste".

Because the United States has not ratified the Basel Convention or its Ban Amendment, and has few domestic federal laws forbidding the export of toxic waste, the Basel Action

Network estimates that about 80% of the electronic waste directed to recycling in the U.S. does not get recycled there at all, but is put on container ships and sent to countries such as China. This figure is disputed as an exaggeration by the EPA, the Institute of Scrap Recycling Industries, and the World Reuse, Repair and Recycling Association.

Independent research by Arizona State University showed that 87–88% of imported used computers did not have a higher value than the best value of the constituent materials they contained, and that "the official trade in end-of-life computers is thus driven by reuse as opposed to recycling".

Trade

Proponents of the trade say growth of internet access is a stronger correlation to trade than poverty. Haiti is poor and closer to the port of New York than southeast Asia, but far more electronic waste is exported from New York to Asia than to Haiti. Thousands of men, women, and children are employed in reuse, refurbishing, repair, and re-manufacturing, unsustainable industries in decline in developed countries. Denying developing nations access to used electronics may deny them sustainable employment, affordable products, and internet access, or force them to deal with even less scrupulous suppliers. In a series of seven articles for The Atlantic, Shanghai-based reporter Adam Minter describes many of these computer repair and scrap separation activities as objectively sustainable.

Opponents of the trade argue that developing countries utilize methods that are more harmful and more wasteful. An expedient and prevalent method is simply to toss equipment onto an open fire, in order to melt plastics and to burn away non-valuable metals. This releases carcinogens and neurotoxins into the air, contributing to an acrid, lingering smog. These noxious fumes include dioxins and furans. Bonfire refuse can be disposed of quickly into drainage ditches or waterways feeding the ocean or local water supplies.

In June 2008, a container of electronic waste, destined from the Port of Oakland in the U.S. to Sanshui District in mainland China, was intercepted in Hong Kong by Greenpeace. Concern over exports of electronic waste were raised in press reports in India, Ghana, Côte d›Ivoire, and Nigeria.

Guiyu

Guiyu in the Shantou region of China is a massive electronic waste processing community. It is often referred to as the "e-waste capital of the world." Traditionally, Guiyu was an agricultural community; however, in the mid-1990s it transformed into an e-waste recycling center involving over 75% of the local households and an additional 100,000 migrant workers. Thousands of individual workshops employ laborers to snip cables, pry chips from circuit boards, grind plastic computer cases into particles, and dip circuit boards in acid baths to dissolve the precious metals. Others work to strip insulation from all wiring in an attempt to salvage tiny amounts of copper wire. Uncontrolled burning, disassembly, and disposal has led to a number of environmental problems such as groundwater contamination, atmospheric pollution, and water pollution either by immediate discharge or from surface runoff (especially near coastal areas), as well as health problems including occupational safety and health effects among those directly and indirectly involved, due to the methods of processing the waste.

A number of studies have been conducted to measure a number of chemicals associated with informal e-waste recycling in the populations. One study enrolled children from Guiyu and a control site 50 km away to measure blood lead levels (BLLs). The average BLL in Guiyu was 15.3 ug/dL compared to 9.9 ug/dL in the control site. In the United States, the CDC has set a reference level for blood lead at 5 ug/dL. High levels of lead in young children can impact IQ and the development of the central nervous system. The highest concentrations of lead were found in the children of parents whose workshop dealt with circuit boards and the lowest was among those who recycled plastic.

Six of the many villages in Guiyu specialize in circuit-board disassembly, seven in plastics and metals reprocessing, and two in wire and cable disassembly. Greenpeace, an environmental group, sampled dust, soil, river sediment and groundwater in Guiyu. They found very high levels of toxic heavy metals and organic contaminants in both places. Lai Yun, a campaigner for the group found "over 10 poisonous metals, such as lead, mercury and cadmium."

Guiyu is only one example of digital dumps but similar places can be found across the world in Nigeria, Ghana, and India. With amounts of e-waste growing rapidly each year urgent solutions are required. While the waste continues to flow into digital dumps like Guiyu there are measures that can help reduce the flow of e-waste.

A suggested preventative step involves the major electronics firms removing the worst chemicals in their products in order to make them safer and easier to recycle.

Other Informal e-Waste Recycling Sites

Guiyu is likely one of the oldest and largest informal e-waste recycling sites in the world, however, there are many sites worldwide, including India, Ghana, Nigeria, and the Philippines. Most research involving informal e-waste recycling has been done in Guiyu, but there are a handful of studies that describe exposure levels in e-waste workers, the community, and the environment.

Bangalore, located in southern India, is often referred as the "Silicon Valley of India" and has a growing informal e-waste recycling sector. Hair samples were collected from workers at an e-waste recycling facility and a e-waste recycling slum community in Bangalore. Levels of V, Cr, Mn, Mo, Sn, Tl, and Pb were significantly higher in the workers at the e-waste recycling facility compared to the e-waste workers in the slum community. However, Co, Ag, Cd, and Hg levels were significantly higher in the slum community workers compared to the facility workers.

A study in Ghana found higher levels of urinary PAH-metabolites in e-waste workers compared to unexposed controls. They also found a greater frequency of complaints of cough, chest pain, and vertigo from those exposed to emissions from the e-waste recycling processes.

Environmental Impact

The processes of dismantling and disposing of electronic waste in developing countries lead to a number of environmental impacts as illustrated in the graphic. Liquid and atmospheric releases end up in bodies of water, groundwater, soil, and air and therefore in land and sea animals – both domesticated and wild, in crops eaten by both animals and human, and in drinking water.

Old keyboards and one mouse.

One study of environmental effects in Guiyu, China found the following:

- Airborne dioxins – one type found at 100 times levels previously measured

- Levels of carcinogens in duck ponds and rice paddies exceeded international standards for agricultural areas and cadmium, copper, nickel, and lead levels in rice paddies were above international standards

- Heavy metals found in road dust – lead over 300 times that of a control village's road dust and copper over 100 times

The Environmental Impact of the Processing of Different Electronic Waste Components

E-Waste Component	Process Used	Potential Environmental Hazard
Cathode ray tubes (used in TVs, computer monitors, ATM, video cameras, and more)	Breaking and removal of yoke, then dumping	Lead, barium and other heavy metals leaching into the ground water and release of toxic phosphor

Printed circuit board (image behind table – a thin plate on which chips and other electronic components are placed)	De-soldering and removal of computer chips; open burning and acid baths to remove metals after chips are removed.	Air emissions and discharge into rivers of glass dust, tin, lead, brominated dioxin, beryllium cadmium, and mercury
Chips and other gold plated components	Chemical stripping using nitric and hydrochloric acid and burning of chips	PAHs, heavy metals, brominated flame retardants discharged directly into rivers acidifying fish and flora. Tin and lead contamination of surface and groundwater. Air emissions of brominated dioxins, heavy metals, and PAHs
Plastics from printers, keyboards, monitors, etc.	Shredding and low temp melting to be reused	Emissions of brominated dioxins, heavy metals and hydrocarbons
Computer wires	Open burning and stripping to remove copper	PAHs released into air, water and soil.

Information Security

E-waste presents a potential security threat to individuals and exporting countries. Hard drives that are not properly erased before the computer is disposed of can be reopened, exposing sensitive information. Credit card numbers, private financial data, account information, and records of online transactions can be accessed by most willing individuals. Organized criminals in Ghana commonly search the drives for information to use in local scams.

Government contracts have been discovered on hard drives found in Agbogbloshie. Multimillion-dollar agreements from United States security institutions such as the Defense Intelligence Agency (DIA), the Transportation Security Administration and Homeland Security have all resurfaced in Agbogbloshie.

E-waste Management

Recycling

Audiovisual components, televisions, VCRs, stereo equipment, mobile phones, other handheld devices, and computer components contain valuable elements and substances suitable for reclamation, including lead, copper, and gold.

One of the major challenges is recycling the printed circuit boards from the electronic wastes. The circuit boards contain such precious metals as gold, silver, platinum, etc. and such base metals as copper, iron, aluminum, etc. One way e-waste is processed is by melting circuit boards, burning cable sheathing to recover copper wire and open-pit acid leaching for separating metals of value. Conventional method employed is mechanical shredding and separation but the recycling efficiency is low. Alternative methods such as cryogenic decomposition have been studied for printed circuit board recycling, and some other methods are still under investigation.

As properly disposing of or reusing electronics can help prevent health problems, reduce greenhouse-gas emissions and create jobs, there have been calls to reform "the methodology for e-waste disposal and re-use in developing countries" with reuse and refurbishing offering a more environmentally friendly and socially conscious alternative to downcycling processes.

Consumer Awareness Efforts

The U.S. Environmental Protection Agency encourages electronic recyclers to become certified by demonstrating to an accredited, independent third party auditor that they meet specific standards to safely recycle and manage electronics. This works to ensure the highest environmental standards are being maintained. Two certifications for electronic recyclers currently exist and are endorsed by the EPA. Customers are encouraged to choose certified electronics recyclers. Responsible electronics recycling reduces environmental and human health impacts, increases the use of reusable and refurbished equipment and reduces energy use while conserving limited resources. The two EPA-endorsed certification programs are: Responsible Recyclers Practices (R2) and E-Stewards. Certified companies ensure they are meeting strict environmental standards which maximize reuse and recycling, minimize exposure to human health or the environment, ensure safe management of materials and require destruction of all data used on electronics. Certified electronics recyclers have demonstrated through audits and other means that they continually meet specific high environmental standards and safely manage used electronics. Once certified, the recycler is held to the particular standard by continual oversight by the independent accredited certifying body. A certification board accredits and oversees certifying bodies to ensure that they meet specific responsibilities and are competent to audit and provide certification.

In the US, the Consumer Electronics Association (CEA) urges consumers to dispose properly of end-of-life electronics through its recycling locator at www.GreenerGadgets.org. This list only includes manufacturer and retailer programs that use the strictest standards and third-party certified recycling locations, to provide consumers assurance that their products will be recycled safely and responsibly. CEA research has found that 58 percent of consumers know where to take their end-of-life electronics, and the electronics industry would very much like to see that level of awareness increase. Consumer electronics manufacturers and retailers sponsor or operate more than 5,000 recycling locations nationwide and have vowed to recycle one billion pounds annually by 2016, a sharp increase from 300 million pounds industry recycled in 2010.

The Sustainable Materials Management Electronic Challenge was created by the United States Environmental Protection Agency (EPA). Participants of the Challenge are manufacturers of electronics and electronic retailers. These companies collect end-of-life (EOL) electronics at various locations and send them to a certified, third-party recycler. Program participants are then able publicly promote and report 100% responsible recycling for their companies.

The Electronics TakeBack Coalition is a campaign aimed at protecting human health and limiting environmental effects where electronics are being produced, used, and discarded. The ETBC aims to place responsibility for disposal of technology products on electronic manufacturers and brand owners, primarily through community promotions and legal enforcement initiatives. It provides recommendations for consumer recycling and a list of recyclers judged environmentally responsible.

The Certified Electronics Recycler program for electronic recyclers is a comprehensive, integrated management system standard that incorporates key operational and continual improvement elements for quality, environmental and health and safety (QEH&S) performance.

The grassroots Silicon Valley Toxics Coalition focuses on promoting human health and addresses environmental justice problems resulting from toxins in technologies.

The World Reuse, Repair, and Recycling Association (wr3a.org) is an organization dedicated to improving the quality of exported electronics, encouraging better recycling standards in importing countries, and improving practices through "Fair Trade" principles.

Take Back My TV is a project of The Electronics TakeBack Coalition and grades television manufacturers to find out which are responsible and which are not.

The e-Waste Association of South Africa (eWASA) has been instrumental in building a network of e-waste recyclers and refurbishers in the country. It continues to drive the sustainable, environmentally sound management of all e-waste in South Africa.

E-Cycling Central is a website from the Electronic Industry Alliance which allows you to search for electronic recycling programs in your state. It lists different recyclers by state to find reuse, recycle, or find donation programs across the country.

Ewasteguide.info is a Switzerland-based website dedicated to improving the e-waste situation in developing and transitioning countries. The site contains news, events, case studies, and more.

StEP: Solving the E-Waste Problem This website of StEP, an initiative founded by various UN organizations to develop strategies to solve the e-waste problem, follows its activities and programs.

Processing Techniques

In many developed countries, electronic waste processing usually first involves dismantling the equipment into various parts (metal frames, power supplies, circuit boards, plastics), often by hand, but increasingly by automated shredding equipment. A typical example is the NADIN electronic waste processing plant in Novi Iskar, Bulgaria—the largest facility of its kind in Eastern Europe. The advantages of this process are the human's ability to recognize and save working and repairable parts, including chips, transistors, RAM, etc. The disadvantage is that the labor is cheapest in countries with the lowest health and safety standards.

Recycling the lead from batteries.

In an alternative bulk system, a hopper conveys material for shredding into an unsophisticated mechanical separator, with screening and granulating machines to separate constituent metal and plastic fractions, which are sold to smelters or plastics recyclers. Such recycling machinery is enclosed and employs a dust collection system. Some of the emissions are caught by scrubbers and screens. Magnets, eddy currents, and Trommel screens are employed to separate glass, plastic, and ferrous and nonferrous metals, which can then be further separated at a smelter.

Leaded glass from CRTs is reused in car batteries, ammunition, and lead wheel weights, or sold to foundries as a fluxing agent in processing raw lead ore. Copper, gold, palladium, silver and tin are valuable metals sold to smelters for recycling. Hazardous smoke and gases are captured, contained and treated to mitigate environmental threat. These methods allow for safe reclamation of all valuable computer construction materials. Hewlett-Packard product recycling solutions manager Renee St. Denis describes its process as: "We move them through giant shredders about 30 feet tall and it shreds everything into pieces about the size of a quarter. Once your disk drive is shredded into pieces about this big, it's hard to get the data off".

An ideal electronic waste recycling plant combines dismantling for component recovery with increased cost-effective processing of bulk electronic waste.

Reuse is an alternative option to recycling because it extends the lifespan of a device. Devices still need eventual recycling, but by allowing others to purchase used electronics, recycling can be postponed and value gained from device use.

Benefits of Recycling

Recycling raw materials from end-of-life electronics is the most effective solution to the growing e-waste problem. Most electronic devices contain a variety of materials, including metals that can be recovered for future uses. By dismantling and providing reuse possibilities, intact natural resources are conserved and air and water pollution caused by hazardous disposal is avoided. Additionally, recycling reduces the amount of greenhouse gas emissions caused by the manufacturing of new products. Another benefit of recycling e-waste is that many of the materials can be recycled and re-used again. Materials that can be recycled include "ferrous (iron-based) and non-ferrous metals, glass, and various types of plastic." "Non-ferrous metals, mainly aluminum and copper can all be re-smelted and re-manufactured. Ferrous metals such as steel and iron can be also be re-used." Due to the recent surge in popularity in 3D printing, certain 3D printers have been designed (FDM variety) to produce waste that can be easily recycled which decreases the amount of harmful pollutants in the atmosphere. The excess plastic from these printers that comes out as a byproduct can also be re-used to create new 3D printed creations.

Benefits of recycling are extended when responsible recycling methods are used. In the U.S., responsible recycling aims to minimize the dangers to human health and the environment that disposed and dismantled electronics can create. Responsible recycling ensures best management practices of the electronics being recycled, worker health and safety, and consideration for the environment locally and abroad. In Europe, metals that are recycled are returned to companies of origin at a reduced cost. Through an extremely committed recycling system, manufacturers in Japan have been pushed to make their products more sustainable. Since many companies were responsible for the recycling of their own products, this imposed responsibility on manufacturers requiring many to re-design their infrastructure. As a result, manufacturers in Japan have the added option to sell the recycled metals.

Electronic Waste Substances

Several sizes of button and coin cell with 2 9v batteries as a size comparison. They are all recycled in many countries since they contain lead, mercury and cadmium.

Some computer components can be reused in assembling new computer products, while others are reduced to metals that can be reused in applications as varied as construction, flatware, and jewelry.

Substances found in large quantities include epoxy resins, fiberglass, PCBs, PVC (polyvinyl chlorides), thermosetting plastics, lead, tin, copper, silicon, beryllium, carbon, iron and aluminium.

Elements found in small amounts include cadmium, mercury, and thallium.

Elements found in trace amounts include americium, antimony, arsenic, barium, bismuth, boron, cobalt, europium, gallium, germanium, gold, indium, lithium, manganese, nickel, niobium, palladium, platinum, rhodium, ruthenium, selenium, silver, tantalum, terbium, thorium, titanium, vanadium, and yttrium.

Almost all electronics contain lead and tin (as solder) and copper (as wire and printed circuit board tracks), though the use of lead-free solder is now spreading rapidly. The following are ordinary applications:

Hazardous

Other health effects

- DNA breaks can increase the likelihood of developing cancer (if the damage is to a tumor suppressor gene)

- DNA damages are a special problem in non-dividing or slowly dividing cells, where unrepaired damages will tend to accumulate over time. On the other hand, in rapidly dividing cells, unrepaired DNA damages that do not kill the cell by blocking replication will tend to cause replication errors and thus mutation

- Elevated Reactive Oxygen Species (ROS) levels can cause damage to cell structures (oxidative stress)

Recyclers in the street in São Paulo, Brazil with old computers

E-Waste Component	Processed Used	Adverse Health Effects
Americium	The radioactive source in smoke alarms.	It is known to be carcinogenic.
Lead	Solder, CRT monitor glass, lead-acid batteries, some formulations of PVC. A typical 15-inch cathode ray tube may contain 1.5 pounds of lead, but other CRTs have been estimated as having up to 8 pounds of lead.	Adverse effects of lead exposure include impaired cognitive function, behavioral disturbances, attention deficits, hyperactivity, conduct problems and lower IQ. These effects are most damaging to children whose developing nervous systems are very susceptible to damage caused by lead, cadmium, and mercury.
Mercury	Found in fluorescent tubes (numerous applications), tilt switches (mechanical doorbells, thermostats), and flat screen monitors.	Health effects include sensory impairment, dermatitis, memory loss, and muscle weakness. Exposure in-utero causes fetal deficits in motor function, attention and verbal domains. Environmental effects in animals include death, reduced fertility, and slower growth and development.
Cadmium	Found in light-sensitive resistors, corrosion-resistant alloys for marine and aviation environments, and nickel-cadmium batteries. The most common form of cadmium is found in Nickel-cadmium rechargeable batteries. These batteries tend to contain between 6 and 18% cadmium. The sale of Nickel-Cadmium batteries has been banned in the European Union except for medical use. When not properly recycled it can leach into the soil, harming microorganisms and disrupting the soil ecosystem. Exposure is caused by proximity to hazardous waste sites and factories and workers in the metal refining industry.	The inhalation of cadmium can cause severe damage to the lungs and is also known to cause kidney damage. Cadmium is also associated with deficits in cognition, learning, behavior, and neuromotor skills in children.
Hexavalent chromium	Used in metal coatings to protect from corrosion.	A known carcinogen after occupational inhalation exposure. There is also evidence of cytotoxic and genotoxic effects of some chemicals, which have been shown to inhibit cell proliferation, cause cell membrane lesion, cause DNA single-strand breaks, and elevate Reactive Oxygen Species (ROS) levels.
Sulphur	Found in lead-acid batteries.	Health effects include liver damage, kidney damage, heart damage, eye and throat irritation. When released into the environment, it can create sulphuric acid.

Brominated Flame Retardants (BFRs)	Used as flame retardants in plastics in most electronics. Includes PBBs, PBDE, DecaBDE, OctaBDE, PentaBDE.	Health effects include impaired development of the nervous system, thyroid problems, liver problems. Environmental effects: similar effects as in animals as humans. PBBs were banned from 1973 to 1977 on. PCBs were banned during the 1980s.
Perfluorooctanoic acid (PFOA)	Used as an antistatic additive in industrial applications and found in electronics, also found in non-stick cookware (PTFE). PFOAs are formed synthetically through environmental degradation.	Studies in mice have found the following health effects: Hepatotoxicity, developmental toxicity, immunotoxicity, hormonal effects and carcinogenic effects. Studies have found increased maternal PFOA levels to be associated with an increased risk of spontaneous abortion (miscarriage) and stillbirth. Increased maternal levels of PFOA are also associated with decreases in mean gestational age (preterm birth), mean birth weight (low birth weight), mean birth length (small for gestational age), and mean APGAR score.
Beryllium oxide	Filler in some thermal interface materials such as thermal grease used on heatsinks for CPUs and power transistors, magnetrons, X-ray-transparent ceramic windows, heat transfer fins in vacuum tubes, and gas lasers.	Occupational exposures associated with lung cancer, other common adverse health effects are beryllium sensitization, chronic beryllium disease, and acute beryllium disease.

Generally Non-hazardous

An iMac G4 that has been repurposed into a lamp (photographed next to a Mac Classic and a flip phone).

E-Waste Component	Process Used
Aluminium	nearly all electronic goods using more than a few watts of power (heatsinks), electrolytic capacitors.

Copper	copper wire, printed circuit board tracks, component leads.
Germanium	1950s–1960s transistorized electronics (bipolar junction transistors).
Gold	connector plating, primarily in computer equipment.
Iron	steel chassis, cases, and fixings.
Lithium	lithium-ion batteries.
Nickel	nickel-cadmium batteries.
Silicon	glass, transistors, ICs, printed circuit boards.
Tin	solder, coatings on component leads.
Zinc	plating for steel parts.

File Deletion

File deletion is a way of removing a file from a computer's file system.

Examples of reasons for deleting files are:

- Freeing the disk space

- Removing duplicate or unnecessary data to avoid confusion

- Making sensitive information unavailable to others

All operating systems include commands for deleting files (rm on Unix, era in CP/M and DR-DOS, del/erase in MS-DOS/PC DOS, DR-DOS, Microsoft Windows etc.). File managers also provide a convenient way of deleting files. Files may be deleted one-by-one, or a whole directory tree may be deleted.

Problem with Accidental Removal

The common problem with deleting files is accidental removal of information that later proves to be important. One way to deal with this is to back up files regularly. Erroneously deleted files may then be found in archives.

Another technique often used is not to delete files instantly, but to move them to a temporary directory whose contents can then be deleted at will. This is how the "recycle bin" or "trash can" works. Microsoft Windows and Apple's Mac OS X, as well as some Linux distributions, all employ this strategy.

In MS-DOS, one can use the undelete command. In MS-DOS the "deleted" files are not really deleted, but only marked as deleted—so they could be undeleted during some time, until the disk blocks they used are eventually taken up by other files. This is how data recovery programs work, by scanning for files that have been marked as deleted. As the space is freed up per byte, rather than per file, this can sometimes cause data to be recovered incompletely. Defragging a drive may prevent undeletion, as the blocks used by deleted file might be overwritten since they are marked as "empty".

Another precautionary measure is to mark important files as read-only. Many operating systems will warn the user trying to delete such files. Where file system permissions exist, users who lack the necessary permissions are only able to delete their own files, preventing the erasure of other people's work or critical system files.

Under Unix-like operating systems, in order to delete a file, one must usually have write permission to the parent directory of that file.

Problem with Sensitive Data

The common problem with sensitive data is that deleted files are not really erased and so may be recovered by interested parties. Most file systems only remove the link to data. But even overwriting parts of the disk with something else or formatting it may not guarantee that the sensitive data is completely unrecoverable. Special software is available that overwrites data, and modern (post-2001) ATA drives include a secure erase command in firmware. However, high security applications and high-security enterprises can sometimes require that a disk drive be physically destroyed to ensure data is not recoverable, as microscopic changes in head alignment and other effects can mean even such measures are not guaranteed.

Retrocomputing

The 1977 Apple II, popular among retrocomputing hobbyists.

Retrocomputing is the use of older computer hardware and software in modern times. Retrocomputing is usually classed as a hobby and recreation rather than a practical application of technology; enthusiasts often collect rare and valuable hardware and software for sentimental reasons. However, some do make use of it. Retrocomputing often starts when a computer user realizes that formerly expensive fantasy systems like IBM mainframes, Digital Equipment Corporation (DEC) superminis, or Silicon Graphics (SGI) workstations have become affordable on the used computer market, usually in a relatively short time after the computers' era of use.

Many hobbyists have personal computer museums, with collections of working vintage computers such as Apple IIs, IBM PCs, ZX Spectrums, Atari, Commodore, Amigas and BBC Micros. Early personal computers based on the S-100 bus are also very popular among collectors, as well as a wide variety of machines running the CP/M operating system, such as Kaypros and Osbornes. However, many users use emulation software on more modern computers rather than using real hardware, in order to enjoy the experience, while preserving the aging electronics of the original. This is not considered to be retrocomputing by some, as it is rather an application of modern computer hardware. A third option is the use of home computer remakes, dedicated appliances, which do the emulation using dedicated hardware.

Retrosystem 2010, a retrocomputing event in Athens

A more serious line of retrocomputing is part of the history of computer hardware. It can be seen as the analogue of experimental archaeology in computing. Some notable examples include the reconstruction of Babbage's Difference engine (more than a century after its design) and the implementation of Plankalkül in 2000 (more than half a century since its inception).

"Homebrew" Computers

Some retrocomputing enthusiasts also consider the 'Homebrewing' (designing and building of retro- and retro-styled computers or kits), to be an important aspect of the hobby, giving new enthusiasts an opportunity to experience more fully what the early years of hobby computing were like. There are several different approaches to this end. Some are exact replicas of older systems, and some are newer designs based on the principles of retrocomputing, while others combine the two, with old and new features in the same package. Examples include:

- Device offered by IMSAI, a modern, updated, yet backward-compatible version and replica of the original IMSAI 8080, one of the most popular early personal systems;

- Several Apple 1 replicas and kits have been sold in limited quantities in recent years, by different builders, such as the "Replica 1", from Briel Computers.;

- A currently ongoing project that uses old technology in a new design is the Z80-based N8VEM;

- The Arduino Retro Computer kit is an open source, open hardware kit you can build and has a BASIC interpreter. There is also a version of the Arduino Retro Computer that can be hooked up to a TV.;

- There is at least one remake of the Commodore 64 using an FPGA configured to emulate the 6502.;

- MSX 2/2+ compatible do-it-yourself kit GR8BIT, designed for the hands-on education in electronics, deliberately employing old and new concepts and devices (high-capacity SRAMs, micro-controllers and FPGA).

Vintage Computers

The personal computer has been around since approximately 1976. But in that time, numerous technological revolutions have left generations of obsolete computing equipment on the junk heap. Nevertheless, in that time, these otherwise useless computers have spawned a sub-culture of vintage computer collectors, who often spend large sums to acquire the rarest of these items, not only to display but restore to their fully functioning glory, including active software development and adaptation to modern uses. This often includes so-called hackers who add-on, update and create hybrid composites from new and old computers for uses for which they were otherwise never intended. Ethernet interfaces have been designed for many vintage 8-bit machines to allow limited connectivity to the Internet; where users can access user groups, bulletin boards and databases of software. Most of this hobby centers on those computers manufactured after 1960, though some collectors specialize in pre-1960 computers as well.

Altair and IMSAI computers with drives

MITS Inc.

Micro Instrumentation and Telemetry Systems (MITS) produced the Altair 8800 in 1975, which is widely regarded as starting the microcomputer revolution.

IMSAI

IMSAI produced a machine similar to the Altair 8800, though considered by many to be a more robust design.

Processor Technology

Processor Technology produced the Sol-20. This was one of the first machines to have a case that included a keyboard; a design feature copied by many of later "home computers".

SWTPC and Altair computers from the 70s

SWTPC

Southwest Technical Products Corporation (SWTPC) produced the SWTPC 6800 and later the SWTPC 6809 kits that employed the Motorola 68xx series microprocessors. The 68xx line was to be followed later by the 6502 processor that was used in many early "home computers", such as the Apple II.

Apple Inc.

The earliest of the Apple Inc. personal computers are among some of the most collectible. They are relatively easy to maintain in an operational state thanks to Apple's use of readily available off-the-shelf parts.

- Apple I: The Apple-1 was Apple's first product and has brought some of the highest prices ever paid for a microcomputer at auction.

- Apple II: The Apple II series of computers are some of the easiest to adapt, thanks to the original expansion architecture designed into them. New peripheral cards are still being designed by an avid thriving community, thanks to the longevity of this platform, manufactured from 1977 through 1993. Numerous websites exist to support not only the legacy users, but new adopters who weren't even born when the Apple II was discontinued by Apple.

- Macintosh: Perhaps because of its friendly design and first commercially successful graphical user interface as well as its enduring Finder application that persists on the most current Macs, the Macintosh is one of the most collected and used of the vintage computers. With dozens of websites around the world, old Macintosh hardware and software is put into daily use. Many maintain vast collections of functional and non-functional systems, which are lovingly maintained and discussed on worldwide user forums. The Macintosh had a strong presence in many early computer labs, creating a strong nostalgia factor for former students who recall their first computing experiences.

Ibm

- The IBM 1130 computing system from 1966 which still has a following of interested users, albeit via an emulator rather than the actual machine.

- The 5100 also has an avid collector and fan base.

- The PC series (5150 PC, 5155 Portable PC, 5160 PC/XT, 5170 PC/AT) has become very popular in recent years, with the earliest models (PC) being considered the most collectible.

Acorn Bbc & Archimedes

- The Acorn BBC Micro was a very popular British computer in the 1980s with home and educational users, and enjoyed near universal usage in British schools into the mid-1990s. It was possible to use 100K 5¼" disks and it had many expansion ports.

- The Archimedes series - the de facto successor to the BBC Micro - has also enjoyed a following in recent years, thanks to its status as one of the first computers to be based around ARM's RISC microprocessor.

Tandy/Radio Shack

- The Tandy/RadioShack Model 100 is still widely collected and used as one of the earliest examples of a truly portable computer. Other Tandy offerings, such as the TRS-80 line, are also very popular, and early systems, like the Model I, in good condition can command premium prices on the vintage computer market.

Sinclair

- The Sinclair ZX81 and ZX Spectrum series were the most popular British home computers of the early 1980s, with a wide choice of emulators available for both platforms. The Spectrum in particular enjoys a cult following due to its popularity as a games platform, with new games titles still being developed even today.

Original "rubber key" Spectrums fetch the highest prices on the second hand market, with the later Amstrad-built models attracting less of a following. The earlier ZX81 is not as popular in original hardware form due to its monochrome display and limited abilities next to the Spectrum, but still unassembled ZX81 kits still appear on eBay occasionally.

Msx

- Although nearly nonexistent in the United States, the MSX architecture has strong communities of fans and hobbyists worldwide, particularly in Japan (where the standard was conceived and developed), Netherlands, Spain, Brazil, Argentina, Russia, Chile, the Middle East and others. New hardware and software are being actively developed to this day as well.

- One of the latest fundamental (from hardware and software perspectives) revivals of the MSX is the GR8BIT.

Robotron

- The Robotron Z1013 was an East German home computer produced by VEB Robotron. It had a U880 processor, 16 kByte RAM and a membrane keyboard.

- The KC 85 series of computers was a modular 8 bit computer system used in East German schools

Internet

There are a number of sites on the Internet catering to vintage computer hobbyists, including web pages, mailing lists, newsgroups, discussion forums, etc. Some are dedicated to certain specific systems while others are more generic and cover many different systems. Erik Klein's Vintage Computer Forum is one example of a discussion page covering all aspects of the hobby.

Some old computers from Commodore International. Amiga 500 (top left), Commodore 128 (top right) and three different variants of the Commodore 64.

cctech, also known as the *"Classic Computers Discussion List"*, is an electronic mailing list about old computer technology, and is run by the Classic Computing organization.

In Popular Culture

In an interview with Conan O'Brien, George R. R. Martin revealed that he writes his books using WordStar 4.0, an application dating back to 1987.

Reception

Retrocomputing (and retrogaming as aspect) has been described by scholars as preservation activity and as aspect of the remix culture.

Retrocomputing Techniques

Home Computer Remake

A home computer remake is a re-creation or re-implementation of classic home computer hardware, usually using updated technology, such as FPGAs.

Description

A remake is a hardware realization, in contrast to an emulator, which is a virtual realization. A remake can also be described as a hardware-based emulator. Some re-makes can function as more than one computer model or architecture. Disputed examples of hardware emulators (which could involve software emulators) are more recent Sega Genesis/Megadrive clones that are cartridge-compatible and can run the games, but use ARM processors as opposed to the Motorola 68k processors of the original Sega Genesis.

Remakes and emulators are a way to keep old software, games, and operating systems alive without having to port them to newer computers or code them again from scratch. Remakes and emulators are methods of digital preservation.

Remakes are not to be confused with hardware clones. Hardware clones are made during a product's initial commercial run, intentionally competing with the original. Remakes are revivals of old, obsolete, or discontinued models. They fill a niche market for retrocomputing researchers, experimenters, hobbyists, gamers, and collectors. Demand for authentic antique hardware often exceeds supply.

Minimalism (Computing)

In computing, minimalism refers to the application of minimalist philosophies and principles in hardware and software design and usage.

History

In the late 1970s and early 1980s programmers had to work within the confines of relatively expensive and limited resources. 8 or 16 kilobytes of RAM was common; 64 kilobytes was considered a vast amount and was the entire address space accessible to the 8-bit CPUs predominant during the earliest generations of personal computers. The most common storage medium was the 5.25 inch floppy disk holding from 88 to 170kB. Hard drives with capacities from 5 to 10 megabytes cost thousands of dollars.

Over time, personal computer memory capacities expanded by orders of magnitude and mainstream programmers took advantage of the added storage to increase their software's capabilities and/or to make development easier by using higher-level languages. By contrast, system requirements for legacy software remained the same. As a result, even the most elaborate, feature-rich programs of yesteryear seem minimalist in comparison with current software. Many of these programs are now considered abandonware. One example of a program whose system requirements once gave it a heavyweight reputation is the GNU Emacs text editor, which gained the backronym "Eight Megabytes And Constantly Swapping" in an era when 8MB was a lot of RAM, but today its mainly textual buffer-based paradigm uses far less resources than desktop metaphor GUI IDEs with comparable features such as Eclipse or Netbeans. In a speech at the 2002 International Lisp Conference, Richard Stallman indicated that minimalism was a concern in his development of GNU and Emacs, based on his experiences with Lisp and system specifications of low-end minicomputers at the time.

As the capabilities and system requirements of common desktop software and operating systems grew throughout the 1980s and 1990s, and as software development became dominated by teams espousing conflicting, faddish software development methodologies, some developers adopted minimalism as a philosophy and chose to limit their programs to a predetermined size or scope. A focus on software optimization can result in minimalist software, as programmers reduce the number of operations their program carries out in order to speed execution.

In the early 21st century, new developments in computing devices have brought minimalism to the forefront. In what has been termed the post-PC era it is no longer necessary to buy a high-end personal computer merely to perform common computing tasks. Mobile computing devices, such as smartphones, tablet computers, netbooks and plug computers, often have smaller memory capacities, less-capable graphics subsystems, and slower processors when compared to the personal computer they are expected to replace. In addition, heavy use of graphics effects like alpha blending drains the battery on these devices faster than a "flat ui". The growing popularity of these stripped-down devices has made minimalism an important design concern. Google's Chrome browser and Chrome OS are often cited as examples of minimalist design. In Windows 8, Microsoft has decided to drop the graphics-intensive Aero user interface in favor of the "simple, squared-off" Metro appearance, which requires less system resources. This change was made in part because of the rise of smaller, battery-powered devices and the need to conserve power. Version 7 of Apple's iOS makes similar changes for user experience reasons.

Usage

Developers may create user interfaces made to be as simple as possible by eliminating buttons and dialog boxes that may potentially confuse the user. Minimalism is sometimes used in its visual arts meaning, particularly in the industrial design of the hardware device or software theme.

Some developers have attempted to create programs to perform a particular function in the fewest lines of code, or smallest compiled executable size possible on a given platform. Some Linux distributions mention minimalism as a goal. Arch Linux, Puppy Linux, Bodhi Linux, CrunchBang Linux, dynebolic and Damn Small Linux are examples. The early development of the Unix system occurred on low-powered hardware, and Dennis Ritchie and Ken Thompson have stated their opinion that this constraint contributed to the system's "elegance of design".

Programming language designers can create minimal programming languages by eschewing syntactic sugar and extensive library functions. Such languages may be Turing tarpits due to not offering standard support for common programming tasks. Creating a minimal Lisp interpreter is a common learning task set before programming students. The Lambda calculus, developed by Alonzo Church defines the most minimal programming language. Scheme, Forth, and Go are cited as examples of minimal programming languages.

The programming hobby of code golf results in minimalist software, but these are typically exercises or code poetry, not usable applications software.

John Millar Carroll, in his book *Minimalism Beyond the Nürnberg Funnel* pointed out that the use of minimalism results in "instant-use" devices such as video games, ATMs, voting machines, and mall kiosks with little-or-no learning curve that do not require the user to read manuals. User Interface researchers have performed experiments suggesting that minimalism, as illustrated by the design principles of parsimony and transparency, bolsters efficiency and learnability. Minimalism is implicit in the Unix philosophies of "everything is a text stream" and "do one thing and do it well".

Abandonware

Abandonware is a product, typically software, ignored by its owner and manufacturer, and for which no product support is available. Although such software is usually still under copyright, the owner may not be tracking or enforcing copyright violations.

In intellectual rights context, abandonware is a software (or hardware) special case of the general concept of *orphan works*.

Definition

Definitions of "abandoned" vary, but in general it is like any item that is abandoned - it is ignored by the owner, and as such product support and possibly copyright enforcement are also "abandoned". It can refer to a product that is no longer available for legal purchase, over the age where the product creator feels an obligation to continue to support it, or where operating systems or hardware platforms have evolved to such a degree that the creator feels continued support cannot be financially justified. In such cases, copyright and support issues are often ignored. Software might also be considered abandoned when it can be used only with obsolete technologies, such as pre-Macintosh Apple computers. A difference between *abandonware* and a *discontinued product* is that the manufacturer has not issued an official notice of discontinuance; instead, the manufacturer is simply ignoring the product.

Abandonware may be computer software or physical devices which are usually computerised in some fashion, such as personal computer games, productivity applications, utility software, or mobile phones.

Types

The term "abandonware" is broad, and encompasses many types of old software.

Commercial software unsupported but still owned by a viable company

> The availability of the software depends on the company's attitude toward the software. In many cases, the company which owns the software rights may not be that which originated it, or may not recognize their ownership. Some companies, such as Borland, make some software available online, in a form of freeware. Others do not make old versions available for free use and do not permit people to copy the software. Many Abandonware websites have been set up to archive and make available copies of unsupported and discontinued operating systems made by Microsoft and Apple, as well as rare development builds of such operating systems that have been leaked by the media or technicians working for said companies - the latter is popular among retrogaming and computing enthusiasts, who value the availability of these operating systems for the ease of access they provide. After Windows XP was discontinued in April 2014, numerous websites started providing the operating system for free, despite the fact that the operating system's usage and popularity prevent it from being considered abandoned.

Commercial software owned by a company no longer in business

> When no owning entity of a software exist, all activities (support, distribution, IP activities etc.) in relationship to this software have ceased. If the rights to a software are non-recoverable in legal limbo ("orphaned work"), also the soft-

ware's rights can't be bought by another company, there can't be copyright enforcement etc. An example of this is Digital Research's original PL/I compiler for DOS, which was considered for many years without owner (now probably owned by Novell).

Shareware whose author still makes it available

Finding historical versions, however, can be difficult since most shareware archives remove past versions with the release of new versions. Authors may or may not make older releases available. Some websites collect and offer for download old versions of shareware, freeware, and (in some cases) commercial applications. In some cases these sites had to remove past versions of software, particularly if the company producing that software still maintains it, or if later software releases introduce digital rights management, whereby old versions could be viewed as DRM circumvention.

Unsupported or unmaintained shareware

In some cases, source code remains available, which can prove a historical artifact. One such case is PC-LISP, still found online, which implements the Franz Lisp dialect. The DOS-based PC-LISP still runs well within emulators and on Microsoft Windows.

Implications

If a software product reaches end-of-life and becomes abandonware, users are confronted with several potential problems: missing purchase availability (besides used software) and missing technical support, e.g. compatibility fixes for newer hardware and operating systems. These problems are exacerbated if software is bound ("dongle") to physical media with a limited life-expectancy (floppy discs, optical media etc.) and backups are impossible because of copy protection or copyright law. If a software is only distributed in a digital, DRM-locked form or as SaaS, the shutdown of the servers will lead to a public loss of the software. If the software product is without alternative, the missing replacement availability becomes a challenge for continued software usage.

Also, once a software product has become abandonware for a developer, even historically important software might get lost forever very easily, as several examples have shown. One of many examples is the closure of Atari in Sunnyvale, California in 1996, when the original source code of several milestones of video game history (like Asteroids or Centipede) was thrown out as trash.

Also, the missing availability of software and the associated source code can be a hindrance for software archeology and research.

Response to Abandonware

Early Abandonware Websites

As response to the missing availability of abandonware, people have distributed old software since shortly after the beginning of personal computing, but the activity remained low-key until the advent of the Internet. While trading old games has taken many names and forms, the term "abandonware" was coined by Peter Ringering in late 1996. Ringering found classic game websites similar to his own, contacted their webmasters, and formed the original *Abandonware Ring* in February 1997. This original webring was little more than a collection of sites linking to adventureclassicgaming.com. Another was a site indexing them all to provide a rudimentary search facility. In October 1997, the Interactive Digital Software Association sent cease and desist letters to all sites within the Abandonware Ring, which led to most shutting down. An unintended consequence (called the Streisand effect in Internet parlance) was that it spurred others to create new abandonware sites and organizations that came to outnumber the original Ring members. Sites formed after the demise of the original Abandonware Ring include Abandonia, Bunny Abandonware and Home of the Underdogs. In later years Abandonware websites actively acquired and received permissions from developers and copyright holders (e.g. Jeff Minter, Magnetic Fields or Gremlin Interactive) for legal redistribution of abandoned works, an example is *World of Spectrum* who acquired the permission from many developers and successfully retracted a DMCA case.

Archives

Several websites archive abandonware for download, including old versions of applications which are difficult to find by any other means. Much of this software fits the definition of "software that is no longer current, but is still of interest", but the line separating the use and distribution of abandonware from copyright infringement is blurry, and the term *abandonware* could be used to distribute software without proper notification of the owner.

The Internet Archive has created an archive of what it describes as "vintage software", as a way to preserve them. The project advocated for an exemption from the United States Digital Millennium Copyright Act to permit them to bypass copy protection, which was approved in 2003 for a period of three years. The exemption was renewed in 2006, and as of 27 October 2009, has been indefinitely extended pending further rulemakings. The Archive does not offer this software for download, as the exemption is solely "for the purpose of preservation or archival reproduction of published digital works by a library or archive." Nevertheless, in 2013 the Internet Archive began to provide antique games as browser-playable emulation via MESS, for instance the Atari 2600 game *E.T. the Extra-Terrestrial*. Since 23 December 2014 the Internet Archive presents via a browser based DOSBox emulation thousands of archived DOS/PC games for *"scholarship and research purposes only"*.

Also the Library of Congress began the long-time preservation of video games with the *Game canon* list around 2006. In September 2012 the collection had nearly 3,000 games from many platforms and also around 1,500 strategy guides. For instance, the source code of the unreleased PlayStation Portable game Duke Nukem: Critical Mass was discovered in August 2014 to be preserved at the Library of Congress.

Since around 2009 the International Center for the History of Electronic Games (ICHEG) has taken a five-pronged approach to video game preservation: original software and hardware, marketing materials and publications, production records, play capture, and finally the source code. In December 2013 the ICHEG received a donation of several SSI video games, for instance Computer Bismarck, including the source code for preservation. In 2014 a collection of Brøderbund games and a "virtually complete" Atari arcade machine source code and asset collection was added.

In 2010 Computer History Museum began with the preservation of source code of important software, beginning with Apple's MacPaint 1.3. In 2012 the APL programming language followed. Adobe Systems, Inc. donated the Photoshop 1.0.1 source code to the collection in February 2013. The source code is made available to the public under an own non-commercial license. On March 25, 2014, Microsoft followed with the donation of MS-DOS variants as well as Word for Windows 1.1a under their own license. On October 21, 2014, Xerox Alto's source code and other resources followed.

In 2012 a group of European museums and organizations started the *European Federation of Game Archives, Museums and Preservation Projects* (EFGAMP) to join forces to *"Preserve Gaming Legacy"*. Also in Japan video game software archival happens since several years.

In 2012 the MOMA started with archiving video games and explicitly tries to get the source code of them.

There are also some cases in which the source code of games was given to a fan community for long-time preservation, e.g. several titles of the *Wing Commander* video game series or Ultima 9 of the Ultima series. In 2008 a hard-drive with all Infocom video game source code appeared from an anonymous source and was archived additionally by the Internet Archive.

Community Support

In response to the missing software support, sometimes the software's user community begins to provide support (bug fixes, compatibility adaptions etc.) even without available source code, internal software documentation and original developer tools. Methods are debugging, reverse engineering of file and data formats, and hacking the binary executables. Often the results are distributed as unofficial patches. Notable examples are *Fallout 2*, *Vampire: The Masquerade – Bloodlines* or even Windows 98. For instance in 2012, when the multiplayer game *Supreme Commander: Forged Alliance* became unsupported abandonware as the official multiplayer server and support was shut down, the game community itself took over with a self-developed multiplayer server and client.

Re-releases by Digital Distribution

With the new possibility of digital distribution arising in mid-2000, the commercial distribution for many old titles became feasible again as deployment and storage costs dropped significantly. A digital distributor specialized in bringing old games out of abandonware is *GOG.com* (formerly called *Good Old Games*) who started in 2008 to search for copyright holders of classic games to release them legally and DRM-free again. For instance, on December 9, 2013 the real-time strategy video game *Conquest: Frontier Wars* was, after ten years of non-availability, re-released by gog.com, also including the source code.

Arguments for and Against Distribution

Proponents of abandonware preservation argue that it is more ethical to make copies of such software than new software that still sells. Those ignorant of copyright law have incorrectly taken this to mean that abandonware is legal to distribute, although no software written since 1964 is old enough for copyright to have expired in the US. Even in cases where the original company no longer exists, the rights usually belong to someone else, though no one may be able to trace actual ownership, including the owners themselves.

Abandonware advocates also frequently cite historical preservation as a reason for trading abandoned software. Older computer media are fragile and prone to rapid deterioration, necessitating transfer of these materials to more modern, stable media and generation of many copies to ensure the software will not simply disappear. Users of still-functional older computer systems argue for the need of abandonware because re-release of software by copyright holders will most likely target modern systems or incompatible media instead, preventing legal purchase of compatible software.

Those who oppose these practices argue that distribution denies the copyright holder potential sales, in the form of re-released titles, official emulation, and so on. Likewise, they argue that if people can acquire an old version of a program for free, they may be less likely to purchase a newer version if the old version meets their needs.

Some game developers showed sympathy for abandonware websites as they preserve their classical game titles.

[...] personally, I think that sites that support these old games are a good thing for both consumers and copyright owners. If the options are (a) having a game be lost forever and (b) having it available on one of these sites, I'd want it to be available. That being said, I believe a game is 'abandoned' only long after it is out of print. And just because a book is out of print does not give me rights to print some for my friends.

—Richard Garriott,

Is it piracy? Yeah, sure. But so what? Most of the game makers aren't living off the revenue from those old games anymore. Most of the creative teams behind all those games have long since left the companies that published them, so there's no way the people who deserve to are still making royalties off them. So go ahead—steal this game! Spread the love!

—Tim Schafer,

If I owned the copyright on Total Annihilation, I would probably allow it to be shared for free by now (four years after it was originally released)

—Chris Taylor,

Law

In most cases, software classed as abandonware is not in the public domain, as it has never had its original copyright officially revoked and some company or individual may still own rights. While sharing of such software is usually considered copyright infringement, in practice copyright holders rarely enforce their abandonware copyrights for a number of reasons – chiefly among which the software is technologically obsolete and therefore has no commercial value, therefore rendering copyright enforcement a pointless enterprise. By default, this may allow the product to de facto lapse into the public domain to such an extent that enforcement becomes impractical.

Rarely has any abandonware case gone to court. But it is still unlawful to distribute copies of old copyrighted software and games, with or without compensation, in any Berne Convention signatory country.

Enforcement of Copyright

Old copyrights are usually left undefended. This can be due to intentional non-enforcement by owners due to software age or obsolescence, but sometimes results from a corporate copyright holder going out of business without explicitly transferring ownership, leaving no one aware of the right to defend the copyright.

Even if the copyright is not defended, copying of such software is still unlawful in most jurisdictions when a copyright is still in effect. Abandonware changes hands on the assumption that the resources required to enforce copyrights outweigh benefits a copyright holder might realize from selling software licenses. Additionally, abandonware proponents argue that distributing software for which there is no one to defend the copyright is morally acceptable, even where unsupported by current law. Companies that have gone out of business without transferring their copyrights are an example of this; many hardware and software companies that developed older systems are long since out of business and precise documentation of the copyrights may not be readily available.

Often the availability of abandonware on the Internet is related to the willingness of copyright holders to defend their copyrights. For example, unencumbered games for Colecovision are markedly easier to find on the Internet than unencumbered games for Mattel Intellivision in large part because there is still a company that sells Intellivision games while no such company exists for the Colecovision.

Dmca

The Digital Millennium Copyright Act (DMCA) can be a problem for the preservation of old software as it prohibits required techniques. In October 2003, the US Congress passed 4 clauses to the DMCA which allow for reverse engineering software in case of preservation.

"3. Computer programs and video games distributed in formats that have become obsolete and which require the original media or hardware as a condition of access. ...The register has concluded that to the extent that libraries and archives wish to make preservation copies of published software and videogames that were distributed in formats that are (either because the physical medium on which they were distributed is no longer in use or because the use of an obsolete operating system is required), such activity is a noninfringing use covered by section 108(c) of the Copyright Act."

—*Exemption to Prohibition on Circumvention of Copyright Protection Systems for Access Control Technologies*

In November 2006 the Library of Congress approved an exemption to the DMCA that permits the cracking of copy protection on software no longer being sold or supported by its copyright holder so that they can be archived and preserved without fear of retribution.

Us Copyright Law

Currently, US copyright law does not recognize the term or concept of "abandonware" while the general concept "orphan works" is recognized. There is a long held concept of abandonment in trademark law as a direct result of the infinite term of trademark protection. Currently, a copyright can be released into the public domain if the owner clearly does so in writing; however this formal process is not considered abandoning, but rather releasing. Those who do not own a copyright cannot merely claim the copyright abandoned and start using protected works without permission of the copyright holder, who could then seek legal remedy.

Hosting and distributing copyrighted software without permission is illegal. Copyright holders, sometimes through the Entertainment Software Association, send cease and desist letters, and some sites have shut down or removed infringing software as a result. However, most of the Association's efforts are devoted to new games, due to those titles possessing the greatest value.

EU Law

In the EU in 2012 an "Orphan Works Directive" (Directive 2012/28/EU) was constituted and is transferred into the member's laws. While the terminology has ambiguities regarding software and especially video games, some scholars argue that Abandonware software video games fall under the definition of *audiovisual works* mentioned there.

Copyright Expiration

Once the copyright on a piece of software has expired, it automatically falls into public domain. Such software can be legally distributed without restrictions. However, due to the length of copyright terms in most countries, this has yet to happen for most software. All countries that observe the Berne Convention enforce copyright ownership for at least 50 years after publication or the author's death. However, individual countries may choose to enforce copyrights for longer periods. In the United States, copyright durations are determined based on authorship. For most published works, the duration is 70 years after the author's death. However, for anonymous works, works published under a pseudonym or works made for hire, the duration is 120 years after publication. In France, copyright durations are 70 years after the relevant date (date of author's death or publication) for either class.

However, because of the length of copyright enforcement in most countries, it is likely that by the time a piece of software defaults to public domain, it will have long become obsolete, irrelevant, or incompatible with any existing hardware. Additionally, due to the relatively short commercial, as well as physical, lifespans of most digital media, it is entirely possible that by the time the copyright expires for a piece of software, it will no longer exist in any form. However, since the largest risk in dealing with abandonware is that of distribution, this may be mitigated somewhat by private users (or organizations such as the Internet Archive) making private copies of such software, which would then be legally redistributable at the time of copyright expiry.

Alternatives to Software Abandoning

There are alternatives for companies with a software product which faces the end-of-life instead of abandoning the software in an unsupported state.

Availability as Freeware

Sometimes user-communities convince companies to voluntarily relinquish copyright on software, putting it into the public domain, or re-license it as free software or as freeware. Transfer of public domain or freely licensed software is perfectly legal, distinguishing it from abandonware which still has full copyright restrictions.

Amstrad is an example which supports emulation and free distribution of CPC and ZX

Spectrum hardware ROMs and software. Borland is another example for a company who released "antique software" as freeware. Smith Engineering permits not-for-profit reproduction and distribution of Vectrex games and documentation.

There are groups that lobby companies to release their software as freeware. These efforts have met with mixed results. One example is the library of educational titles released by MECC. MECC was sold to Brøderbund, which was sold to The Learning Company. When TLC was contacted about releasing classic MECC titles as freeware, the documentation proving that TLC held the rights to these titles could not be located, and therefore the rights for these titles are "in limbo" and may never be legally released. That the copyright situation of vintage out-of-print software is lost or unclear is not uncommon.

Support by Source Code Release

The problem of missing technical support for a software can be most effectively solved when the source code becomes available. Therefore, several companies decided to release the source code specifically to allow the user communities to provide further technical software support (bug fixes, compatibility adaptions etc.) themselves, e.g. by community patches or source ports to new computing platforms. For instance, in December 2015 Microsoft released the Windows Live Writer source code to allow the community to continue the support.

Id Software and 3D Realms are early proponents in this practice, releasing the source code for the game engines of some older titles under a free software license (but not the actual game content, such as levels or textures). Also Falcon 4.0's lead designer Kevin Klemmick argued in 2011 that availability of the source code of his software for the community was a good thing:

I honestly think this [source code release] should be standard procedure for companies that decide not to continue to support a code base.

— Kevin Klemmick, interviewed by Bertolone, Giorgio (2011-03-12). "Interview with Kevin Klemmick - Lead Software Engineer for Falcon 4.0". Cleared-To-Engage. Archived from the original on 2011-03-18. Retrieved 2014-08-31.

The chilling effect of drawing a possible lawsuit can discourage release of source code. Efforts to persuade IBM to release OS/2 as open source software were ignored since some of the code was co-developed by Microsoft.

Nevertheless, several notable examples of successfully opened commercial software exist, for instance, the web browser Netscape Communicator, which was released by Netscape Communications on March 31, 1998. The development was continued under the umbrella of the Mozilla Foundation and Netscape Communicator became the basis of several browsers, such as Mozilla Firefox.

Another important example for open sourced general software is the office suite StarOffice which was released by Sun Microsystems in October 2000 as OpenOffice.org and is in continued development as LibreOffice and Apache OpenOffice.

There are also many examples in the video game domain: Revolution Software released their game *Beneath a Steel Sky* as freeware and gave the engine's source code to the authors of ScummVM to add support for the game. Other examples are *Myth II, Call to Power II* and Microsoft's *Allegiance* which were released to allow the community to continue the support.

Greening

Greening is the process of transforming artifacts such as a space, a lifestyle or a brand image into a more environmentally friendly version (i.e. 'greening your home' or 'greening your office'). The act of greening involves incorporating "green" products and processes into one's environment, such as the home, work place, and general lifestyle.

Green Qualities

These "green" qualities include, but are not limited to:

- reduced toxicity
- re-usability
- energy efficiency
- responsible packaging and labelling
- recycled content
- intelligent design
- responsible manufacturing techniques
- reduction of personal environmental hazards
- alleviate heat island effect

Assessment

Environmentally friendly companies such as Green Home have developed a rigorous approval policy that allows consumers to qualify each product based upon these criteria as they apply to specific product categories.

Greening Methods

Community Greens

Community Greens, sometimes referred to as *backyard commons*, *urban commons*, or pocket neighborhoods, are shared open green spaces on the inside of city blocks, created either when residents merge backyard space or reclaim underutilized urban land such as vacant lots and alleyways. These shared spaces are communally used and managed only by the residents whose homes abut them. They are not a public park, a private backyard, or a community garden; however, they can function as all three.

Community Greens

Community Greens is an organization concerned with the development of shared green spaces in residential neighborhoods in American cities. These green spaces are community greens. The Community Greens movement believes that such an approach presents the best opportunity to add usable green space to American cities, by converting under utilized backyards and dysfunctional alleys into functional and beautiful shared green spaces that are owned, managed, and enjoyed by the people who live around them.

This has led communities in numerous American cities, including Boston, Sacramento, Baltimore, New York, and San Francisco, taking down their backyard fences, to create backyard commons.

Community Green behind the Luzerne-Glover Block

Community Greens are multi-functional spaces for gardening, recreation, and leisure which are designed to provide social, economic, and environmental benefits to urban residents. The creation of backyard commons can lead to an increased interaction with neighbors throughout the planning and implementation process, which may result in a stronger over-all sense of community. Other possible social benefits that are claimed include decreased crime, from having more eyes on the street, and safe places where children can play and adults relax. Community Greens, like other types of urban green spaces, can significantly improve the ecological functioning of urban habitats. Vegetation and permeable pavement can slow storm water runoff and increase groundwater, which in turn can reduce pollutant flowing into nearby bodies of water during a storm. Urban environments are often significantly warmer than outlying suburbs, mostly due to the prevalence of heat retentive concrete surfaces. City trees can mitigate this effect through shading, etc.. The taking down of backyards fences transforms fragmented habitats to connected corridors for urban wildlife. Furthermore because city dwellers recognize value in green space, often simply from an aesthetic standpoint, and this may be reflected through increased property values.

The Beginnings in Maryland

- Community Greens was founded as an offshoot of Ashoka, by Ashoka founder Bill Drayton in 2001.

- Spearheaded a state wide legislature in Maryland, which changed the City Charter which allows the city to lease and gate alleys to homeowners.

- Started the Alley Gating and Alley Greening pilot project in Baltimore in 2006.

- Local ordinance passed in the City of Baltimore in 2007.

- As of 2009, there are 3 completely gated and greened alleys in Baltimore and over 70 applications from interested neighborhoods.

Baltimore: The Pilot Project

In 2002, a group of residents from the Patterson Park neighborhood approached the Patterson Park Community Development Corporation (CDC) looking for a way to improve the dirty, crime-ridden alley that ran behind their homes. Simultaneously, Community Greens also approached the Patterson Park CDC looking for an alley they could use as a pilot project in Baltimore, and these two groups were put into contact. The Luzerne-Glover block was granted a temporary permit from the city to gate their alleyway, despite the fact that it was not yet legal to gate a right-of-way. Community Greens, the Patterson Park CDC, the Patterson Park Neighborhood Association and the Luzurne-Glover block group turned to University of Maryland law professor Barbara Bezdek, who enlisted the help of her law students, who researched existing laws and current uses of Baltimore alleyways.

Looking for a permanent solution, these groups aligned themselves with the Mayor's Department of Neighborhoods and took their case to Annapolis. Maryland Delegate Peter A. Hammen sponsored House Bill 1533 to amend the Baltimore City Charter, allowing the City to close alleyways and lease them to interested parties. The Luzerne-Glover group testified before the Environmental Committee of the state legislature and House Bill 1533 passed in 2004. Community Greens then worked with law firm Hogan & Hartson, in collaboration with Barbara Bezdek, to draft a city alley gating and leasing ordinance, which was then submitted to the City of Baltimore.

In April 2007, the Baltimore City Council under the leadership of newly elected Mayor Shelia Dixon, passed the Gating and Greening Alleys ordinance, enabling Baltimore residents to legally gate and green the alleys behind their homes, contingent on a requirement that 80% of the homeowners on the alley submit consent forms and 100% of property owners to approve projects that impede traffic flow. Projects requiring 100% consent include removal of existing concrete and installation of permeable pavement and large tree plantings; beautification and smaller greening projects that do not impede traffic only require 80% approval.

The Luzerne-Glover Block alley before gating and greening

Interested communities need approval from several Baltimore City Departments, including Solid Waste, Transportation, Fire, and Police to ensure that proposed alley projects meet the necessary infrastructural conditions. Once initial approval is received, residents submit an application to the Department of Public Works, which includes the necessary consent forms from homeowners and a signed affidavit stating that attempts were made to obtain approval from all homeowners.

The Luzerne Glover neighborhood now has their green alleyway, and is the first community in Baltimore to use the Gating and Greening Alleys ordinance. So far, seven Baltimore communities have successfully completed this process and have created eight useful Community Greens from the underutilized alleyways in their backyards.

Other Existing Greens

Montgomery Park - Boston, Massachusetts

Montgomery Park is a third of an acre common space shared by 85 households in Boston's South End neighborhood. From the time the neighborhood was first developed in the 1860s through the 1970s, the park was separated from the residents' backyards by an alley and fence. In more recent years, residents began to connect the park to their individual yards by replacing the alley separating their private backyards from the park with a narrower brick walkway and tearing down backyard fences. Additionally, residents have moved garbage collection to the front street and convinced utility companies to bury service lines. Residents now enjoy greater access to the shared Community Green, and use it for gardening and recreation.

Stoney Creek Apartments - Livermore, California

Stoney Creek Apartments were constructed in 1992 to address a lack of affordable housing in Livermore, California. Ten residential buildings share five courtyards, accessible from a small alleyway. Homes have front and back patios in addition to the shared courtyards, and neighbors have the opportunity to enjoy this shared outdoor space and experience a heightened sense of community. The inward-facing position of these patios allows parents to keep a watchful eye on children playing in the courtyard.

Chandlers Yard - Baltimore, Maryland

Located in the Federal Hill neighborhood of downtown Baltimore and hidden behind eleven narrow rowhouses, Chandler's Yard is a tree-shaded courtyard that was carved out of the backyards of the surrounding homes by developer Bill Struever who wanted to make living on the block more attractive to potential buyers. Struever convinced some of the property owners in this block to give up part of their backyards for this shared courtyard and pay a small construction fee. The result is a beautiful shared space for these residents, who enjoy a heightened sense of security and increased property values.

Village Homes - Davis, California

The Village Homes development is situated on a sixty-acre parcel in suburban Davis, California. While home lots are smaller than the average for Davis, clusters of eight homes share common green spaces accessible from private backyards. In addition to these courtyards, Village Homes residents share two large parks, two vineyards, and numerous small orchards and community gardens. Gardens are irrigated by naturally-flowing creeks, which also serve as natural filters that eliminate the need for an expensive storm-water sewer system. Residents enjoy environmental and aesthetic benefits from their shared spaces, and the increased sense of community is evident; Village Homes residents on average know 40 neighbors, compared to an average of 17 acquaintances reported in a nearby traditional suburban development.

Jackson Heights - Queens, New York

For nearly nine decades, the historic neighborhood of Jackson Heights in Queens has maintained its fourteen block-long shared interior courtyards, and is considered to be the first "garden apartments" constructed in the United States. These spaces have helped to sustain the blocks' distinctive appeal since their development in the early decades of the last century. Over the years, Jackson Heights residents fought to hold on to their gardens and green spaces in the midst of a city where high property values create an intense pressure to develop any available open space. The preservation of these shared spaces has increased both community pride and residential property values; if these inner courtyards had been developed, Jackson Heights property values would have dropped by one-third. In 1993, Jackson Heights was officially declared a historic district by the NYC Landmark Preservation Commission, furthering sense of place for this community.

St. Francis Square - San Francisco, California

A cooperative apartment community, St. Francis Square consists of 299 apartments wrapped around three shared, open spaces. These community greens are full of gardens, basketball courts, playgrounds, and quiet places to relax. St. Francis Square was established as a limited equity co-operative apartment, although it converted to market rate ownership in 2004.

The Hope Community - Minneapolis, Mn

Fifteen years ago, this community was situated in an area of Minneapolis plagued by urban flight and crime. Instrumental to the rehabilitation and renovation of this neighborhood is a local community development corporation, Hope Community, Inc. In the late 1980s, residents were selling their properties for one dollar because they had lost all value; this is when the Hope Community began purchasing houses in the hopes of developing affordable rental units clustered together on a single block. Today, nine rehabilitated houses abut community areas, a playground, and gardens. Property values are steadily increasing and the community itself is stabilizing. By following principles of Crime Prevention Through Environmental Design (CPTED) and creating defensible space, the Hope Community has created a sense of shared ownership and community in a troubled urban area.

Green Alleys

Something similar has taken place in various cities in North America, involving the greening of back lanes or alleys. This includes Chicago, Seattle, Los Angeles, Washington, D.C., and Montréal, Canada, who have started to reclaim their alleys from garbage and crime by greening the service lanes, or back ways, that run behind some houses.

Chicago, Illinois has about 1,900 miles (3,100 km) of alleyways. In 2007, the Chicago Department of Transportation started converting conventional alleys which were paved with asphalt into so called Green Alleys. This program, called the Green Alley

Program, is supposed to enable easier water runoff, as the alleyways in Chicago are not connected to the sewer system. With this program, the water will be able to seep through semi-permeable concrete or asphalt in which a colony of fungi and bacteria will establish itself. The bacteria will help breakup oils before the water is absorbed into the ground. The lighter color of the pavement will also reflect more light, making the area next to the alley cooler. The greening of such alleys or laneways can also involve the planting of native plants to further absorb rain water and moderate temperature.

Green Alliance

Green Alliance is a charity and an independent think tank focused on ambitious leadership for the environment in the United Kingdom (UK). Founded in 1979, it works with influential leaders from the NGO, business, and political communities. Its work aims to generate new thinking and dialogue, and has increased political action and support for environmental solutions in the UK.

The organisation has three main subject themes on which it conducts analysis and makes policy recommendations. These are: Natural Environment, Low Carbon Energy and Resource Stewardship. Many of Green Alliance's projects involve collaboration with partner businesses and NGOs to ensure policies are both progressive and workable.

Since the early 1990s Green Alliance has convened the chief executives of UK's foremost environmental non-governmental organisations (NGOs) in strategic alliances, leading joint advocacy to government on major policy issues.

Notable individuals involved in Green Alliance's early years include Maurice Ash, Tim Beaumont and Tom Burke. The current director Matthew Spencer was previously head of government affairs at the Carbon Trust. He sits on the Department for Energy and Climate Change's Carbon Capture and Storage Development Forum and regularly advises government on energy and environment policy.

Green Alliance's staff is based in central London. It has diverse support base, including an individual membership of approximately 400 leading environment professionals, and many partnerships with leading businesses and NGOs.

Green Alliance's blog is a platform for insight and commentary from both staff and external opinion formers on UK environmental policy and politics. It was a finalist in the 'Green and Eco' category at the UK Blog Awards 2016.

In 2009 Green Alliance was awarded 'Think Thank of the Year' at the Public Affairs News Awards. In 2016 it was awarded 'Best Environmental Campaign by an NGO' at the Green Ribbon Political Awards for its work in securing the party leaders' climate agreement and received Highly Commended for NGO of the Year at the Business Green Leaders Awards 2016.

History

Green Alliance was founded in March 1979 to inject an 'ecological perspective into the political life of Britain'. Founder Maurice Ash said at the launch "We're a bunch of optimists. We're not the doomsters. We believe in the possibilities of the future..." Ever since, Green Alliance has focused on influencing UK government environment policy.

Early on, in the 1980s, it elicited the first ever environmental policy statements from all the main UK political parties. It was the first organisation in the UK to raise genetic modification (GM) as an environmental issue at the end of the 80s, with its risk analysis which informed new government regulation.

IN 1998, it hosted the Green Globe Network of foreign policy experts, supported by cross-cutting funding from three government departments: the Foreign and Commonwealth Office, Department for International Development and the Department for Environment, Food and Rural Affairs.

A number of new organisations and initiatives have arisen from Green Alliance's work over the years, including the National Food Alliance (now Sustain), the Environment Agency, the Carbon Trust and the UK's Renewables Obligation.

An independent evaluation of Green Alliance's work in 2008 concluded: "Green Alliance's key strengths are its influence within Westminster and Whitehall and its understanding of the political process. Green Alliance's relationships with politicians, special advisers and civil servants are its primary strength. It is considered influential within government. Those in government value the ability of Green Alliance to bring them policy positions and mediate relations with NGOs."

Examples of its influence include securing new government strategies on green taxes and microgeneration, increases to landfill tax and more investment in recycling, brokering a historic party leaders' pledge on climate change, helping to win the argument in government to end unabated coal and proposing a new approach to support for renewables into the 2020s which was adopted by the government.

Expert Opinion

Green Alliance staff and associates are often quoted as expert sources in national and specialist news outlets such as *The Guardian, The Independent, BusinessGreen, The Financial Times* and *The ENDS Report.* Subjects which have recently been commented on by Green Alliance experts in major news outlets are; The Green Investment Bank; London's environmental policies and the fifth carbon budget.

Community Gardening

Strathcona Heights Community Garden in Ottawa, Canada

A community garden is a single piece of land gardened collectively by a group of people.

Purpose

Crops at the former South Central Farm in Los Angeles, California, United States

According to Marin Master Gardeners, "a community garden is any piece of land gardened by a group of people, utilizing either individual or shared plots on private or public land". Community gardens provide fresh products and plants as well as contributing to a sense of community and connection to the environment and an opportunity for satisfying labor and neighborhood improvement. They are publicly functioning in terms of ownership, access, and management, as well as typically owned in trust by local governments or not for profit associations.

Community gardens vary widely throughout the world. In North America, community gardens range from familiar "victory garden" areas where people grow small plots of vegetables, to large "greening" projects to preserve natural areas, to tiny street beautification planters on urban street corners. Some grow only flowers, others are nurtured communally and their bounty shared. There are even non-profits in many major cities that offer assistance to low-income families, children groups, and community organizations by helping them develop and grow their own gardens. In the UK and the rest of Europe, closely related "allotment gardens" can have dozens of plots, each measuring hundreds of square meters and rented by the same family for generations. In the developing world, commonly held land for small gardens is a familiar part of the landscape, even in urban areas, where they may function as market gardens. They also practice crop rotations with versatile plants such as peanuts, tomatoes and much more.

Community gardens may help alleviate one effect of climate change, which is expected to cause a global decline in agricultural output, making fresh produce increasingly unaffordable.Community gardens are also an increasingly popular method of changing the built environment in order to promote health and wellness in the face of urbanization. The built environment has a wide-range of positive and negative effects on the people who work, live, and play in a given area, including a person's chance of developing obesity Community gardens encourage an urban community's food security, allowing citizens to grow their own food or for others to donate what they have grown. Advocates say locally grown food decreases a community's reliance on fossil fuels for transport of food from large agricultural areas and reduces a society's overall use of fossil fuels to drive in agricultural machinery. A 2012 op-ed by community garden advocate Les Kishler examines how community gardening can reinforce the so-called "positive" ideas and activities of the Occupy movement.

Community gardens improve users' health through increased fresh vegetable consumption and providing a venue for exercise.A fundamental part of good health is a diet rich in fresh fruits, vegetables, and other plant based foods. Community gardens provide access to such foods for the communities in which they are located. Community gardens are especially important in communities with large concentrations of low socioeconomic populations, as a lack fresh fruit and vegetable availability plagues these communities at disproportionate rates.

The gardens also combat two forms of alienation that plague modern urban life, by bringing urban gardeners closer in touch with the source of their food, and by breaking down isolation by creating a social community. Community gardens provide other social benefits, such as the sharing of food production knowledge with the wider community and safer living spaces. Active communities experience less crime and vandalism.

Ownership

Land for a community garden can be publicly or privately held. One strong tradition in North American community gardening in urban areas is cleaning up abandoned vacant lots and turning them into productive gardens. Alternatively, community gardens can be seen as a health or recreational amenity and included in public parks, similar to ball fields or playgrounds. Historically, community gardens have also served to provide food during wartime or periods of economic depression. Access to land and security of land tenure remains a major challenge for community gardeners and their supporters throughout the world, since in most cases the gardeners themselves do not own or control the land directly.

Mobility Community gardening

Some gardens are grown collectively, with everyone working together; others are split into clearly divided plots, each managed by a different gardener (or group or family). Many community gardens have both "common areas" with shared upkeep and individual/family plots. Though communal areas are successful in some cases, in others there is a tragedy of the commons, which results in uneven workload on participants, and sometimes demoralization, neglect, and abandonment of the communal model. Some relate this to the largely unsuccessful history of collective farming.

Unlike public parks, whether community gardens are open to the general public is dependent upon the lease agreements with the management body of the park and the community garden membership. Open or closed-gate policies vary from garden to garden. However, in a key difference, community gardens are managed and maintained with the active participation of the gardeners themselves, rather than tended only by a professional staff. A second difference is food production: Unlike parks, where plantings are ornamental (or more recently ecological), community gardens often encourage food production by providing gardeners a place to grow vegetables and other crops. To facilitate this, a community garden may be divided into individual plots or tended in a communal fashion, depending on the size and quality of a garden and the members involved.

Plot Size

In Britain, the 1922 Allotment act specifies "an allotment not exceeding 40 [square] poles in extent"; since a rod, pole or perch is 5.5 yards in length, 40 square rods is 1210 square yards or 10890 square feet (equivalent to a large plot of 90 ft x 121 ft). In practice, plot sizes vary; Lewisham offers plots with an "average size" of "125 metres square".

In America, plots vary; for example, plots of 3m × 6m (10 ft × 20 ft = 200 square feet) and 3m x 4.5m (10 ft x 15 ft) are listed in Alaska. Montgomery Parks in Maryland lists plots of 200, 300, 400 and 625 square feet. In Canada, plots of 20x20 and 10x10 feet, as well as smaller "raised beds", are listed in Vancouver.

Location

The location of a community garden is a critical factor in how often the community garden is used and who visits it. Exposure to a community garden is much more likely for an individual if they are able to walk or drive to the location, as opposed to public transportation.The length of travel time is also a factor. Those who live within a 15 minute or less travel distance are more likely to visit a community garden as compared to those with a longer travel time. Such statistics should be taken into consideration when choosing a location for a community garden for a target population.

Plant Choice and Physical Layout

While food production is central to many community and allotment gardens, not all have vegetables as a main focus. Restoration of natural areas and native plant gardens are also popular, as are "art" gardens. Many gardens have several different planting elements, and combine plots with such projects as small orchards, herbs and butterfly gardens. Individual plots can become "virtual" backyards, each highly diverse, creating a "quilt" of flowers, vegetables and folk art.

Raised beds are sometimes used to separate plants from surrounding natural plants or to improve ease of access, especially for disabled gardeners.

Group and Leadership Selection

The community gardening movement in North American prides itself on being inclusive, diverse, pro-democracy, and supportive of community involvement. Gardeners may be of any cultural background, young or old, new gardeners or seasoned growers, rich or poor. A garden may have only a few people active, or hundreds.

Finally, all community gardens have a structure. The organization depends in part on whether the garden is "top down" or "grassroots". There are many different organizational models in use for community gardens. Some elect boards in a democratic fashion, while others can be run by appointed officials. Some are managed by non-profit organizations, such as a community gardening association, a community association, a church, or other land-owner; others by a city's recreation or parks department, a school or University.

Gardeners may form a grassroots group to initiate the garden, such as the Green Guerrillas of New York City, or a garden may be organized "top down" by a municipal agency. The Los Gatos, California-based non-profit Community Gardens as Appleseeds offers free assistance in starting up new community gardens around the world.

Membership Rules and Fees

In most cases, gardeners are expected to pay annual dues to help with garden upkeep, and the organization must manage these fees. The tasks in a community garden are many, including upkeep, mulching paths, recruiting members, and fund raising. Rules and an 'operations manual' are both invaluable tools, and ideas for both are available at the ACGA.

Health Effects of Community Gardens

Community gardens have been shown to have positive health effects on those who participate in the programs, particularly in the areas of decreasing body mass index and lower rates of obesity. Studies have found that community gardens in schools have been found to improve average body mass index in children. A 2013 study found that 17% of obese or overweight children improved their body mass index over seven weeks. Specifically, 13% of the obese children achieved a lower body mass index in the overweight range, while 23% of overweight children achieved a normal body mass index. Many studies have been performed largely in low-income, Hispanic/Latino communities in the United States. In these programs, gardening lessons were accompanied by nutrition and cooking classes and optional parent engagement. Successful programs highlighted the necessity of culturally tailored programming.

There is some evidence to suggest that community gardens have a similar effect in adults. A study found that community gardeners in Utah had a lower body mass index than their non-gardening siblings and unrelated neighbors. Administrative records were used to compare body mass indexes of community gardeners to that of unrelated neighbors, siblings, and spouses. Gardeners were less likely to be overweight or obese than their neighbors, and gardeners had lower body mass indexes than their siblings. However, there was no difference in body mass index between gardeners and their spouses which may suggest that community gardening creates healthy habits for the entire household.

Participation in a community garden has been shown to increase both availability and consumption of fruits and vegetables in households. A study showed an average increase in availability of 2.55 fruits and 4.3 vegetables with participation in a community garden. It also showed that children in participating households consumed an average of two additional servings per week of fruits and 4.9 additional servings per week of vegetables.

Policy Implications

There is strong support among American adults for local and state policies and policy changes that support community gardens. A study found that 47.2 % of American adults supported such policies. However, community gardens compete with the interests of developers. Community gardens are largely impacted and governed by policies at the city level. In particular, zoning laws strongly impact the possibility of community gardens. The momentum for rezoning often comes from the public need for access to fruits and vegetables. Rezoning is necessary in many cities for a parcel of land to be designated a community garden, but rezoning doesn't guarantee garden will not be developed in the future.

Further policies can be enacted to protect community gardens from future development. For example, New York State reached a settlement in 2002 which protected hundreds of community gardens which had been established by the Parks and Recreation Department GreenThumb Program from future development.

At times, zoning policy lags behind the development of community gardens. In these cases, community gardens may exist illegally. Such was the case in Detroit when hundreds of community gardens were created in abandoned spaces around the city. The city of Detroit created agricultural zones in 2013 in the middle of urban areas to legitimize the over 355 "illegal" community gardens.

Examples

Australia

The first Australian community garden was established in 1977 in Nunawading, Victoria followed soon after by Ringwood Community Garden in March 1980.

Spain

The squatted social center Can Masdeu is home to one of the largest community gardens in Barcelona.

Most older Spaniards grew up in the countryside and moved to the city to find work. Strong family ties often keep them from retiring to the countryside, and so urban community gardens are in great demand. Potlucks and paellas are common, as well as regular meetings to manage the affairs of the garden.

United Kingdom

In the United Kingdom, community gardening is generally distinct from allotment gardening, though the distinction is sometimes blurred. Allotments are generally plots of land rented to individuals for their cultivation by local authorities or other public bodies—the upkeep of the land is usually the responsibility of the individual plot owners. Allotments tend (but not invariably) to be situated around the outskirts of built-up areas. Use of allotment areas as open space or play areas is generally discouraged. However, there are an increasing number of community-managed allotments, which may include allotment plots and a community garden area, many of them overseen by the Federation of City Farms and Community Gardens (a registered charity).

The community garden movement is of more recent provenance than allotment gardening, with many such gardens built on patches of derelict land, waste ground or land owned by the local authority or a private landlord that is not being used for any purpose. A community garden in the United Kingdom tends to be situated in a built-up area and is typically run by people from the local community as an independent, non-profit organisation (though this may be wholly or partly funded by public money).

It is also likely to perform a dual function as an open space or play area (in which role it may also be known as a 'city park') and—while it may offer plots to individual cultivators—the organisation that administers the garden will normally have a great deal of the responsibility for its planting, landscaping and upkeep. An example inner-city garden of this sort is Islington's Culpeper Community Garden, which is a registered charity, or Camden's Phoenix Garden.

United States

Taiwan

There is an extensive network of community gardens and collective urban farms in Taipei City often occupying areas of the city that are waiting for development. Flood-prone river banks and other areas unsuitable for urban construction often become legal or illegal community gardens. The network of the community gardens of Taipei are referred to as *Taipei organic acupuncture* of the industrial city.

Mali

Often externally supported, community gardens become increasingly important in developing countries, such as West African Mali to bridge the gap between supply and requirements for micro-nutrients and at the same time strengthen an inclusive development.

Village Green

A village green is a common open area within a village or other settlement. Traditionally, a village green was often common grassland at the centre of an agricultural or other rural settlement, and was used for grazing. Some also have a pond, often originally for watering stock such as cattle.

The village green also provided, and may still provide, an open-air meeting place for the local people, which may be used for public celebrations such as May Day festivities.

The term *village green* evokes a grassy rural environment. However the term is used more broadly to encompass woodland, moorland, sports grounds, and even—in part—

buildings and roads. The green may also be positioned away from the centre of the village, especially if the village has moved, or been absorbed into a larger settlement.

A village green in Cambridgeshire, England. Benches face a large pond - a common sight in many village greens.

Distribution

Some historical village greens have been lost as a result of the agricultural revolution and urban development. Greens are now most likely to be found in the older villages of mainland Europe, the United Kingdom, and older areas of the United States. Town expansion in the mid-20th century led in England to the formation of local conservation societies, often centring on village green preservation, as celebrated and parodied in The Kinks' album *The Kinks Are The Village Green Preservation Society*. The Open Spaces Society is a present-day UK national campaigning body which continues this movement.

A large green in the village of Pritzhagen, Germany

The term can also apply to urban parks. In the United States, the most famous example of a town green is probably the New Haven Green in New Haven, Connecticut. New Haven was founded by settlers from England and was the first planned city in the United States. Originally used for grazing livestock, the Green dates from the 1630s and is well preserved today despite lying at the heart of the city centre. The largest green in the U.S. is a mile in length, and can be found in Lebanon, Connecticut. One of the most unusual is the Dartmouth Green in Hanover, New Hampshire, which was owned and cleared by the college in 1770. The college, not the town, still owns it and surrounded it with buildings as a sort of collegiate quadrangle in the 1930s, although its origin as a town green remains apparent.

A fine example of a traditional American town green exists in downtown Morristown, NJ. The Morristown Green dates from 1715 and has hosted events ranging from executions to clothing drives.

Town and Village Greens

A house on a village green in Cumbria, England

The village green in Stanford in the Vale, Oxfordshire, England

Apart from the general use of the term, *village green* has a specific legal meaning in England and Wales, and also includes the less common term *town green*. Town and village greens were defined in the Commons Registration Act 1965, as amended by the Countryside and Rights of Way Act 2000, as land:

1. which has been allotted by or under any Act for the exercise or recreation of the inhabitants of any locality

2. or on which the inhabitants of any locality have a customary right to indulge in lawful sports and pastimes

3. or if it is land on which for not fewer than twenty years a significant number of the inhabitants of any locality, or of any neighbourhood within a locality, have indulged in lawful sports and pastimes as of right.

Registered greens are now governed by the Commons Act 2006, but the fundamental test of whether land is a town and village green remains the same. Thus land can become a village green if it has been used for twenty years without force, secrecy or request (*nec vi, nec clam, nec precario*). Village green legislation is often used to try to frustrate development. Recent case law (*Oxfordshire County Council vs Oxford City Council and Robinson*) makes it clear that registration as a green would render any development which prevented continuing use of the green as a criminal activity under the Inclosure Act 1857 and the Commons Act 1876. This leads to some most curious areas being claimed as village greens, sometimes with success. Recent examples include a bandstand, two lakes and a beach.

The Open Spaces Society states that in 2005 there were about 3,650 registered greens in England covering 8,150 acres (3,298 ha) and about 220 in Wales covering about 620 acres (251 ha).

Examples

A notable example of a village green is that in the village of Finchingfield in Essex, England, which is said to be "the most photographed village in England". The green dominates the village, and slopes down to a duck pond, and is occasionally flooded after heavy rain. The small village of Car Colston in Nottinghamshire, England, has two village greens, totaling 29 acres (12 ha).

A village green in Zuidlaren, in the Netherlands

The village green in Willerzie, Belgium

Some greens that used to be a common or otherwise at the centre of a village have been swallowed up by a city growing around them. Sometimes they become a city park or a square, and manage to maintain a sense of place. London has several of these: Newington Green, originally a Dissenting village, is one good example, with its church anchoring its north end.

The village green in Gretton, Northamptonshire - one of the few remaining British villages to retain its stocks and whipping post.

There are two places in the United States called Village Green: Village Green-Green Ridge, Pennsylvania, and Village Green, New York. Some New England towns, along with some areas settled by New Englanders such as the townships in the Connecticut Western Reserve, refer to their town square as a village green. The only village green in the United States still used for agriculture lies in Lebanon, Connecticut. This green is also one of the largest in the nation.

In Indonesia, especially in Java, a similar place is called *Alun-Alun*. It is a central part of Javanese village architecture and culture.

The northern part of the province of Drenthe in the Netherlands is also known for its village greens. Zuidlaren is the village with the largest number of village greens in the Netherlands.

The Błonia Park, originally established in the Middle Ages, is an example of a large village green in Kraków, Poland.

Park

A park is an area of natural, semi-natural, or planted space set aside for human enjoyment and recreation or for the protection of wildlife or natural habitats. It may consist of grassy areas, rocks, soil, and trees, but may also contain buildings and other artifacts such as monuments, fountains or playground structures. In North America, many parks have fields for playing sports such as soccer, baseball and football, and paved areas for games such as basketball. Many parks have trails for walking, biking and other activities. Some parks are built adjacent to bodies of water or watercourses, and these parks may comprise a beach or boat dock area. Often, the smallest parks are in urban areas, where a park may take up only a city block or less. Urban parks often have benches for sitting and they may contain picnic tables and barbecue grills. Parks have differing rules regarding whether dogs can be brought into the park: some parks prohibit dogs; some parks allow them with restrictions (e.g., use of a leash); and some parks, which may be called "dog parks," permit dogs to run off-leash.

The largest parks can be vast natural areas of hundreds of thousands of square kilometres (thousands of square miles), with abundant wildlife and natural features such as mountains and rivers. In many large parks, camping in tents is allowed with a permit. Many natural parks are protected by law, and users may have to follow restrictions (e.g., rules against open fires or bringing in glass bottles). Large national and sub-national parks are typically overseen by a park ranger or a park warden. Large parks may have areas for canoeing and hiking in the warmer months and, in some northern hemisphere countries, cross-country skiing and snowshoeing in colder months.

History

Depiction of a medieval hunting park from a 15th-century manuscript

The first parks were English deer parks, land set aside for hunting by royalty and the aristocracy in medieval times. They had walls or thick hedges around them to keep game animals (e.g., stags) in and people out. It was strictly forbidden for commoners to hunt animals in these deer parks.

These game preserves evolved into landscaped parks set around mansions and country houses from the sixteenth century onwards. These may have served as hunting grounds but they also proclaimed the owner's wealth and status. An aesthetic of landscape design began in these stately home parks where the natural landscape was enhanced by landscape architects such as Capability Brown. As cities became crowded, the private hunting grounds became places for the public.

With the Industrial revolution parks took on a new meaning as areas set aside to preserve a sense of nature in the cities and towns. Sporting activity came to be a major use for these urban parks. Areas of outstanding natural beauty were also set aside as national parks to prevent their being spoiled by uncontrolled development.

In some parks or time periods with high pollen counts, parks tend to be avoided.

Design

Park design is influenced by the intended purpose and audience, as well as by the available land features. A park intended to provide recreation for children may include a playground. A park primarily intended for adults may feature walking paths and decorative landscaping. Specific features, such as riding trails, may be included to support specific activities.

The design of a park may determine who is willing to use it. Walkers may feel unsafe on a mixed-use path that is dominated by fast-moving cyclists or horses. Different landscaping and infrastructure may even affect children's rates of use of parks according to gender. Redesigns of two parks in Vienna suggested that the creation of multiple semi-enclosed play areas in a park could encourage equal use by boys and girls.

Parks are part of the urban infrastructure: for physical activity for families and communities to gather and socialize, or for a simple respite. Research reveals that people who exercise outdoors in green-space derive greater mental health benefits. Providing activities for all ages, abilities and income levels is important for the physical and mental well-being of the public.

Design for Safety

A well-lit path in Dehli's Garden of Five Senses

Parks need to feel safe for people to use them. Research shows that perception of safety can be more significant in influencing human behaviour than actual crime statistics. If citizens perceive a park as unsafe, they might not make use of it at all.

There are a number of features that contribute to whether or not a park feels safe. Elements in the physical design of a park, such as an open and welcoming entry, good visibility (sight lines), and appropriate lighting and signage can all make a difference. Regular park maintenance, as well as programming and community involvement can also contribute to a feeling of safety.

Women as a Measure of Safety

In the United States, the standard for safety in parks is increasingly measured by whether women feel safe in that particular location. This was originally identified by the urban sociologist William H Whyte in his studies decades ago in New York. Research reveals that women have a different sense of safety compared to men, whether they are walking in their neighborhood or in a park. Dan Biederman, President of the Bryant Park Corp. stated "Women pick up on visual cues of disorder better than men do.... And if women don't see other women, they tend to leave." Whether or not a woman feels safe can determine how much physical activity she has and if it will reach the recommended level for good health and disease prevention. Park designers and planners can take several steps to increase women's safety from sexual assault or other assault, including providing sufficient lighting, having patrols by police officers or other safety officials, and providing emergency buttons for summoning assistance.

Active and Passive Recreation Areas

Parks can be divided into active and passive recreation areas. Active recreation is that which has an urban character and requires intensive development. It often involves cooperative or team activity, including playgrounds, ball fields, swimming pools, gymnasiums, and skateparks. Active recreation such as team sports, due to the need to provide substantial space to congregate, typically involves intensive management, maintenance, and high costs. Passive recreation, also called "low intensity recreation" is that which emphasizes the open-space aspect of a park and allows for the preservation of natural habitat. It usually involves a low level of development, such as rustic picnic areas, benches and trails.

Many smaller *neighborhood parks* are receiving increased attention and valuation as significant community assets and places of refuge in heavily populated urban areas. Neighborhood groups around the world are joining together to support local parks that have suffered from urban decay and government neglect.

Passive recreation typically requires little management and can be provided at very low costs. Some open space managers provide nothing other than trails for physical activity in the form of walking, running, horse riding, mountain biking, snow shoeing, or cross-country skiing; or sedentary activity such as observing nature, bird watching, painting, photography, or picnicking. Limiting park or open space use to passive recreation over all or a portion of the park's area eliminates or reduces the burden of managing active recreation facilities and developed infrastructure. Many ski resorts combine active recreation facilities (ski lifts, gondolas, terrain parks, downhill runs, and lodges) with passive recreation facilities (cross-country ski trails).

Parks Owned or Operated by Government

National Parks

A national park is a reserve of land, usually, but not always declared and owned by a national government, protected from most human development and pollution. Although this may be so, it is not likely that the government of a specific area owns it, rather the community itself. National parks are a protected area of International Union for Conservation of Nature Category II. This implies that they are wilderness areas, but unlike pure nature reserves, they are established with the expectation of a certain degree of human visitation and supporting infrastructure.

Northeast Greenland National Park, the world's largest national park

While this type of national park had been proposed previously, the United States established the first "public park or pleasuring-ground for the benefit and enjoyment of the people", Yellowstone National Park, in 1872, although Yellowstone was not gazetted as a national park. The first officially designated national park was Mackinac Island, gazetted in 1875. Australia's Royal National Park, established in 1879, was the world's second officially established national park.

The largest national park in the world is the Northeast Greenland National Park, which was established in 1974 and currently protects 972,001 km² (375,000 sq mi)

Sub-national Parks

In some Federal systems, many parks are managed by the sub-national levels of government. In Brazil, the United States, and some states in Mexico, as well as in the Australian state of Victoria, these are known as state parks, whereas in Argentina, Canada and South Korea, they are known as provincial or territorial parks.

Urban Parks

Yoyogi Park is a large urban park in Tokyo.

A park is an area of open space provided for recreational use, usually owned and maintained by a local government. Parks commonly resemble savannas or open woodlands, the types of landscape that human beings find most relaxing. Grass is typically kept short to discourage insect pests and to allow for the enjoyment of picnics and sporting activities. Trees are chosen for their beauty and to provide shade.

Sad Janka Kráľa park in Bratislava (Slovakia)

Some early parks include the la Alameda de Hércules, in Seville, a promenaded public mall, urban garden and park built in 1574, within the historic center of Seville; the City Park, in Budapest, Hungary, which was property of the Batthyány family and was later made public.

An early purpose built public park was Derby Arboretum which was opened in 1840 by Joseph Strutt for the mill workers and people of the city. This was closely followed by Princes Park in the Liverpool suburb of Toxteth, laid out to the designs of Joseph Paxton from 1842 and opened in 1843. The land on which the Princes park was built was purchased by Richard Vaughan Yates, an iron merchant and philanthropist, in 1841 for £50,000. The creation of Princes Park showed great foresight and introduced a number of highly influential ideas. First and foremost was the provision of open space for the benefit of townspeople and local residents within an area that was being rapidly built up. Secondly it took the concept of the designed landscape as a setting for the suburban domicile, an idea pioneered by John Nash at Regent's Park, and re-fashioned it for the provincial town in a most original way. Nash's remodeling of St James's Park from 1827 and the sequence of processional routes he created to link The Mall with Regent's Park completely transformed the appearance of London's West End. With the establishment of Princes Park in 1842, Joseph Paxton did something similar for the benefit of a provincial town, albeit one of international stature by virtue of its flourishing mercantile contingent. Liverpool had a burgeoning presence on the scene of global maritime trade before 1800 and during the Victorian era its wealth rivalled that of London itself.

The form and layout of Paxton's ornamental grounds, structured about an informal lake within the confines of a serpentine carriageway, put in place the essential elements of his much imitated design for Birkenhead Park. The latter was commenced in 1843 with the help of public finance and deployed the ideas he pioneered at Princes Park on a more expansive scale. Frederick Law Olmsted visited Birkenhead Park in 1850 and praised its qualities. Indeed, Paxton is widely credited as having been one of the principal influences on Olmsted and Calvert's design for New York's Central Park of 1857.

Central Park in New York City is the most-visited urban park in the U.S.

Another early public park is the Peel Park, Salford, England opened on August 22, 1846. Another possible claimant for status as the world's first public park is Boston Common (Boston, Massachusetts, USA), set aside in 1634, whose first recreational promenade, Tremont Mall, dates from 1728. True park status for the entire common seems to have emerged no later than 1830, when the grazing of cows was ended and renaming the Common as Washington Park was proposed (renaming the bordering Sentry Street to Park Street in 1808 already acknowledged the reality).

Linear Parks

A linear park is a park that has a much greater length than width. A typical example of a linear park is a section of a former railway that has been converted into a park called a rail trail or greenway (i.e. the tracks removed, vegetation allowed to grow back). Parks are sometimes made out of oddly shaped areas of land, much like the vacant lots that often become city neighborhood parks. Linked parks may form a greenbelt.

Country Parks

In some countries, especially the United Kingdom, *country parks* are areas designated for recreation, and managed by local authorities. They are often located near urban populations, but they provide recreational facilities typical of the countryside rather than the town.

Private Parks

Private parks are owned by individuals or businesses and are used at the discretion of the owner. There are a few types of private parks, and some which once were privately maintained and used have now been made open to the public.

Hunting parks were originally areas maintained as open space where residences, industry and farming were not allowed, often originally so that nobility might have a place to hunt. These were known for instance, as *deer parks* (deer being originally a term meaning any wild animal). Many country houses in Great Britain and Ireland still have parks of this sort, which since the 18th century have often been landscaped for aesthetic effect. They are usually a mixture of open grassland with scattered trees and sections of woodland, and are often enclosed by a high wall. The area immediately around the house is the garden. In some cases this will also feature sweeping lawns and scattered trees; the basic difference between a *country house's park* and its garden is that the park is grazed by animals, but they are excluded from the garden.

Common Land

Common land is land owned collectively by a number of persons, or by one person, but over which other people have certain traditional rights, such as to allow their livestock to graze upon it, to collect firewood, or to cut turf for fuel.

A person who has a right in, or over, common land jointly with another or others is called a commoner.

This article deals mainly with common land in England, Wales and Scotland. Although the extent is much reduced due to enclosure of common land from the millions of acres that existed until the 17th century, a considerable amount of common land still exists, particularly in upland areas, and there are over 7,000 registered commons in England alone.

Common land or former common land is usually referred to as a common; for instance, Clapham Common or Mungrisdale Common.

Origins

Originally in medieval England the common was an integral part of the manor, and was thus part of the estate held by the lord of the manor under a feudal grant from the Crown or a superior peer, who in turn held his land from the Crown which owned all land. This manorial system, founded on feudalism, granted rights of land use to different classes. These would be *appurtenant* rights, that is the ownership of rights belonged to tenancies of particular plots of land held within a manor. A commoner would be the person who, for the time being, was the occupier of a particular plot of land. Some rights of common were said to be *in gross*, that is, they were unconnected with tenure of land. This was more usual in regions where commons are more extensive, such as in the high ground of Northern England or on the Fens, but also included many village greens across England and Wales. Most land with appurtenant commons rights is adjacent to the common or even surrounded by it, but in a few cases it may be some considerable distance away.

Modern-day pannage, or common of mast, in the New Forest

Conjectural map of a mediaeval English manor. The part allocated to "common pasture" is shown in the north-east section, shaded green.

Historically Manorial courts defined the details of many of the rights of common allowed to manorial tenants, and such rights formed part of the copyhold tenancy whose terms were defined in the manorial court roll.

Example rights of common are:

- Pasture. Right to pasture cattle, horses, sheep or other animals on the common land. The most widespread right.

- Piscary. Right to fish.

- Turbary. Right to take sods of turf for fuel.

- Common in the Soil This is a general term used for rights to extract minerals such as sands, gravels, marl, walling stone and lime from common land.

- Mast or pannage. Right to turn out pigs for a period in autumn to eat mast (beech mast, acorns and other nuts).

- Estovers. Right to take sufficient wood for the commoner's house or holding; usually limited to smaller trees, bushes (such as gorse) and fallen branches.

On most commons, rights of pasture and pannage for each commoner are tightly defined by number and type of animal, and by the time of year when certain rights could be exercised. For example, the occupier of a particular cottage might be allowed to graze fifteen cattle, four horses, ponies or donkeys, and fifty geese, whilst the numbers allowed for their neighbours would probably be different. On some commons (such as the New Forest and adjoining commons), the rights are not limited by numbers, and instead a *marking fee* is paid each year for each animal *turned out*. However, if excessive use was made of the common, for example, in overgrazing, a common would be *stinted*, that is, a limit would be put on the number of animals each commoner was allowed to graze. These regulations were responsive to demographic and economic pressure. Thus rather than let a common become degraded, access was restricted even further.

Types of Common

Snake's head fritillary, North Meadow, Cricklade. This is grazed as Lammas common land.

View of the Scafell massif from Yewbarrow, Wasdale, Cumbria. In the valley are older enclosures and higher up on the fell-side are the parliamentary enclosures following straight lines regardless of terrain.

Pasture Commons

Pasture commons are those where the primary right is to pasture livestock. In the uplands, they are largely moorland, on the coast they may be salt marsh, sand dunes or cliffs, and on inland lowlands they may be downland, grassland, heathland or wood pasture, depending on the soil and history. These habitats are often of very high nature conservation value, because of their very long continuity of management extending in some cases over many hundreds of years. In the past, most pasture commons would have been grazed by mixtures of cattle, sheep and ponies (often also geese). The modern survival of grazing on pasture commons over the past century is uneven.

Arable and Haymeadow Commons

Surviving commons are almost all pasture, but in earlier times, arable farming and haymaking were significant, with strips of land in the common arable fields and common haymeadows assigned annually by lot. When not in use for those purposes, such commons were grazed. Examples include the common arable fields around the village of Laxton in Nottinghamshire, and a common meadow at North Meadow, Cricklade.

Lammas Rights

Lammas rights entitled commoners to pasture following the harvest, between Lammas day, 12 August (N.S.), to 6 April, even if they did not have other rights to the land. Such rights sometimes had the effect of preventing enclosure and building development on agricultural land.

Enclosure and Decline

Most of the medieval common land of England was lost due to enclosure. In English social and economic history, enclosure or inclosure is the process which ends traditional rights such as mowing meadows for hay, or grazing livestock on common land formerly held in the open field system. Once enclosed, these uses of the land become restricted to the owner, and it ceases to be land for the use of commoners. In England and Wales the term is also used for the process that ended the ancient system of arable farming in open fields. Under enclosure, such land is fenced (*enclosed*) and *deeded* or *entitled* to one or more owners. The process of enclosure began to be a widespread feature of the English agricultural landscape during the 16th century. By the 19th century, unenclosed commons had become largely restricted to large areas of rough pasture in mountainous areas and to relatively small residual parcels of land in the lowlands.

Enclosure could be accomplished by buying the ground rights and all common rights to accomplish exclusive rights of use, which increased the value of the land. The other method was by passing laws causing or forcing enclosure, such as Parliamentary enclosure. The latter process of enclosure was sometimes accompanied by force, resistance, and bloodshed, and remains among the most controversial areas of agricultural and economic history in England.

Enclosure is considered one of the causes of the British Agricultural Revolution. Enclosed land was under control of the farmer who was free to adopt better farming practices. There was widespread agreement in contemporary accounts that profit making opportunities were better with enclosed land. Following enclosure, crop yields and livestock output increased while at the same time productivity increased enough to create a surplus of labour. The increased labour supply is considered one of the factors facilitating the Industrial Revolution.

Following the era of enclosure, there was relatively little common land remaining of value. Some residual commoners remained, until such as after the Second World War, lowland commons became neglected because commoners could find better-paid work in other sectors of the economy. As a result, they largely stopped exercising their rights, and relatively few commoners exist today.

Modern Use

Much common land is still used for its original purpose. The right to graze domestic stock is by far the most extensive commoners right registered, and its ongoing use contributes significantly to agricultural and rural economies. Rights to graze sheep are registered on 53% of the Welsh and 16% of the English commons. Cattle are registered on 35% of Welsh and 20% of English commons, whilst horses and ponies are registered on 27% of Welsh and 13% of English commons. In some cases rights to graze goats, geese and ducks are registered, whilst in others the type of livestock is not specified. These figures relate to the number of common land units, and due to discrepancies in the registers and large numbers of small commons with no rights in England, the apparent distinction between Wales and England may be exaggerated.

Today, despite the diverse legal and historical origins of commons, they are managed through a community of users, comprising those who hold rights together with the owner(s) of the soil. Such communities generally require joint working to integrate all interests, with formal or informal controls and collaborative understandings, often coupled with strong social traditions and local identity.

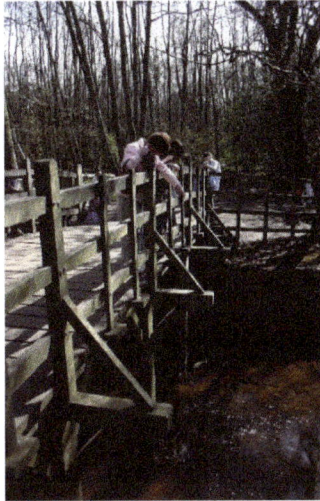

Poohsticks Bridge in Ashdown Forest, an area of common land.

However, 26% of commons in Wales, and as many as 65% in England, have no common rights shown on the registers. Such areas are derived from wastes of manors, where rights probably existed formerly. When such open habitats are no longer grazed they revert to scrub and then dense woodland, losing the grassy or heathland vegetation which may have occupied the land continuously for many centuries. In 2007 Ashdown Forest, the Sussex heathland which was the setting for the Winnie-the-Pooh stories, became the centre of a dispute between some local residents and the Forest's governing body, the Board of Conservators, which is responsible for administering the Forest's 2,400 hectares (5,900 acres) of common land. The Conservators wished to restore the Forest's landscape to one that predominantly consisted of heathland—its defining characteristic until the mid-twentieth century, but something that was in danger of being lost after the Second World War as a result of the advance of woodland into traditional heathland areas when, as one commentator stated..

...returning soldiers gave up trying to scratch a living out of the forest. Whereas once hundreds of commoners used the wood and heath—their livestock obliging by chewing down young tree shoots—today there is only one commercial grazer.

The Conservators were forced to intervene to stem the invasion of trees, scrub and bracken that threatened the ecologically precious heathlands, cutting down saplings, removing scrub and mowing the bracken. Some residents complained that the results looked like a First World War battle field. This is not a problem restricted to this common, but according to Jonathan Brown writing in the Independent on 21 April 2007 "similar debates are raging between locals and the authorities at other heathland areas in the New Forest and Surrey".

In 2008 the Foundation for Common Land was created in the UK to try to enhance the understanding and protection of commons.

Law of Common Land - England and Wales

The legal position concerning common land has been confused, but recent legislation has sought to remedy this and remove the legal uncertainties so that commons can be better used and protected.

Most commons are based on ancient rights, that is to say common law (coincidence of term only), which pre-date statute law laid down by parliament. The exact rights which apply to individual commons were in some cases documented but more often were based on long-held traditions. A major reform was started in 1965 with a national register of common land which recorded the land ownership and the rights of any commoners, and two other significant acts have followed.

Owners of land in general have all the rights of exclusive ownership, to use the land as they wish. However, for common land the owner's rights are restricted, and other people known as commoners have certain rights over the land. The landowner may retain other rights to the land, such as rights to minerals and large timber, and to any common rights left unexercised by the commoners. The commoners will continue to exercise their rights, or have a document which describes their rights, which may be part of the deeds of another property. A number of commoners still exercise rights, for example, there are 500 practising commoners in the New Forest, and there is a federation of commoners in Cumbria. In many cases commons have no existing commoners, the rights having been neglected.

Erection of Cottages Act 1588

There was a belief that if an Englishman or woman could build a house on common land, raise the roof over their head and have a fire in the hearth between sunrise and sunset, then they could have the right of undisturbed possession. The belief was actually a fallacy, but to stop landless peasants unlawfully squatting on commons, an act known as the Erection of Cottages Act 1588 (31 Eliz c. 7, long title "An Act against the erecting and maintaining of Cottages"), was introduced.

Commons Act 1876

Under the Commons Act 1876 some 36 commons in England and Wales were regulated. The act also enabled the confirmation of Orders providing for the inclosure of common land or common fields.

Commons Act 1899

The Commons Act 1899 provides a mechanism of enabling district councils and National Park authorities to manage commons where their use for exercise and recreation is the prime consideration and where the owner and commoners do not require a direct voice in the management, or where the owner cannot be found. There are at least 200 schemes of management made under the 1899 act.

The Law of Property Act 1925

The Law of Property Act 1925, which still forms the core of English property law, has two provisions for common land:

- Section 193 gave the right of the public to "air and exercise" on Metropolitan commons and those in pre-1974 urban districts and boroughs. This constituted about one fifth of the commons, but the 1925 Act did not give this right to commons in essentially rural areas (although some urban districts had remarkably rural extent, such as the Lakes Urban District), which had to wait for the 2000 CROW Act.

- Section 194 restricted the inclosure of commons, which would now require Ministerial consent.

Commons Registration Act 1965

The UK government regularised the definitions of common land with the Commons Registration Act 1965, which established a register of common land.

Not all commons have owners, but all common land by definition is registered under 1965 Commons registration Act, along with the rights of any commoners if they still exist. The registration authorities are the County Councils, and when there is no ownership, a local council, such as a parish council is normally given guardianship by vesting the property under the Act (section 8).

An online database of registered common land was compiled by DEFRA in 1992-93 as part of a survey of the condition and wildlife of commons. The official up to date Registers of common land are held by the Commons Registration Authorities.

The following registration information is held:

- Land Section

This includes a description of the land, who applied to register the land, and when the land became finally registered. There are also related plans which show the boundaries of the land.

- Rights Section

This includes a description of the rights of common (e.g. a right to graze a certain amount of sheep), the area of common over which the right is exercisable, the name of the holder of the right and whether the right is attached to land in the ownership of the holder of the right (the commoner) or is a right held in gross i.e. unattached to land.

- Ownership Section

This includes details of the owner(s) of the common land. Entries in this section however, are not held to be conclusive.

Unfortunately, numerous inconsistencies and irregularities remained, mainly because a period of only 3 years was given for registration submissions. However, there is there is now an opportunity to clear these up under the 2006 Act, and to add land omitted under the 1965 act.

Countryside and Rights of Way Act 2000 (CROW)

Other than for those commons covered by the Law of Property Act 1925, the Commons Act 1899 and certain other statutes, the public did not have the right to use or enjoy common land if they were not a commoner. However, the Countryside and Rights of Way Act 2000 gave the public the Freedom to roam freely on all registered common land in England and Wales. The new rights were introduced region by region through England and Wales, with completion in 2005. Maps showing accessible areas have been produced, and are available online as "open access maps" produced by Natural England. Commons are included in the public access land now shown on the Ordnance Survey Explorer Maps.

Commons Act 2006

The Commons Act 2006 is an important new piece of legislation.

The Act:

- Enables commons to be managed more sustainably by commoners and land-owners working together through commons councils with powers to regulate grazing and other agricultural activities

- Provides better protection for common land and greens - this includes reinforcing existing protections against abuse, encroachment and unauthorised development

- Recognises that the protection of common land has to be proportionate to the harm caused and that some specified works can be carried out without the need for consent

- Requires commons registration authorities to bring their registers up-to-date by recording past changes affecting the registers during a 'transitional period', and to keep the registers up-to-date by recording new changes affecting the registers - commons registration authorities will have new powers to correct many of the mistakes in the registers

- Sets out new, clearer criteria for the registration of town or village greens

- Prohibits the severance of rights of common grazing, preventing commoners from selling, leasing or letting their rights away from the property to which rights are attached, though temporary severance of such rights is permitted for renewable terms of up to two years (in England) and five years (in Wales).

Several hundred square kilometres of 'waste land' that was provisionally registered under the Commons Registration Act 1965 was not, in fact, finally registered. As a consequence, it ceased to be recognised as common land. A partial remedy for this defect in the earlier legislation is provided by the Commons Act 2006. Under Schedule 2(4) to the Act, applications that failed to achieve final registration under the 1965 Act may, in certain circumstances, be reconsidered – offering, in effect, a second chance for the land to be confirmed ('re-registered') as common. Land that is re-registered in this way will enjoy the special legal protection afforded to common land. It will also become subject in due course to the public right of access introduced by the Countryside and Rights of Way Act 2000; or depending on location, may qualify as a section 193 'urban' common (in which case, it would also be subject to a right of access for horse-riders).

Fencing

The windmill on Wimbledon Common.

The act of transferring resources from the commons to purely private ownership is known as *enclosure*, or (especially in formal use, and in place names) *Inclosure*. The Inclosure Acts were a series of private Acts of Parliament, mainly from about 1750 to 1850, which enclosed large areas of common, especially the arable and haymeadow land and the better pasture land.

The maintenance of fences around a common is the responsibility of the occupiers of the adjacent enclosed land, not (as it would be with enclosed land) the responsibility of the owners of the grazed livestock. This can lead to difficulties where not all adjacent occupiers maintain their fences properly. However the fencing of land within a registered common is not allowed, as this is a form of enclosure and denies use of the land to others.

A celebrated landmark case of unauthorised fencing of a common was in 1866 by Lord Brownlow who illegally enclosed 434 acres of Berkhamsted Common to add to his Ashridge Estate. Brownlow had failed to buy out the commoners, so resorted to this action. A public outcry followed, and the Commons Preservation Society found a champion in Augustus Smith who had the inclination and the money to act, and himself held commons rights. Smith hired 120 navvies armed with hammers, chisels and crowbars, who on the night of

6 March 1866, under the aegis of the newly formed Commons Preservation Society (now the Open Spaces Society), felled to the ground two miles of iron railings. Soon after, local people flocked in. Lord Brownlow took action against Augustus Smith and the court case lasted until 1870 when it ended with the complete vindication of Smith.

Controls on Development

Development of common land is strictly controlled. The government states that common land should be open and accessible to the public, and the law restricts the kind of works that can be carried out on commons. HM Planning Inspectorate is responsible for determining applications under the 2006 Act regarding common land in England, and several other pieces of legislation regarding commons and greens. All applications are determined on behalf of the Secretary of State for the Environment, Food and Rural Affairs (Defra).

Under section 38 of the Commons Act 2006, you need consent to carry out any restricted works on land registered as common land under the Commons Registration Act 1965.Restricted works are any that prevent or impede access to or over the land. They include fencing, buildings, structures, ditches, trenches, embankments and other works, where the effect of those works is to prevent or impede access. They also include, in every case, new solid surfaces, such as for a new car park or access road.

Boards of Conservators and Commons Councils

Some commons are managed by Boards of Conservators for the wider public benefit. However, for areas where these are not established, or an improved system is required, the Commons Act 2006 provides for the establishment of Commons Councils to manage common land.

The Standard Constitution Regulations relating to commons councils were formally approved in April 2010, and Commons Councils are most likely to be useful where they can improve current management practices. This may be where commons are in agricultural use, but where it can be difficult to reach agreement on collective management. Commons Councils are voluntary and can be established only where there is substantial support among those with interests in the land, such as; the Commoners (especially those who actively exercise their rights); owners and other legal interests.

Commons Councils enable decisions to be made by majority voting, so relieving the burden of trying to reach unanimous decisions. They will have the power to make rules about agricultural activities, the management of vegetation, and the exercise of common rights, which are binding on all those with interests on a common.

Roadways

Commons are often crossed by unfenced public roads, and this leads to another problem on modern pasture commons where grazing survives (or is to be reintroduced). Historically, the roads would have been cart-tracks, and there would have been no conflict between their horse-drawn (or ox-drawn) traffic and the pastured animals, and no great difficulty if pastured animals wandered off the common along the roads. However, these roads now have fast motorised traffic which does not mix safely with animals. To continue (or restore) grazing, such roads may need fencing or at least blocking at the edge of the common with cattle grids — however fencing a common is reminiscent of the process of enclosure, historically fatal to its survival, and permission for fencing on a common is a strictly controlled process within the UK planning system.

A parliamentary enclosure road near Lazonby in Cumbria. The roads were made as straight as possible, and the boundaries much wider than a cart width to reduce the ground damage of driving sheep and cattle.

Public roads through enclosed common land were made to an accepted width between boundaries. In the late eighteenth century this was at least 60 feet (18 m), but from the 1790s this was decreased to 40 feet (12 m), and later 30 feet (9.1 m) as the normal maximum width. The reason for these wide roads to was to prevent excessive churning of the road bed, and allow easy movement of flocks and herds of animals.

Scotland

Commoning has probably existed in Scotland for over a millennium. However, there is no modern legislation relating to commons which formally identifies the extent of common land or clarifies the full range of rights. The right of turbary – the ability to cut peat as fuel – clearly exists in large parts of Scotland, whilst the scale of such rights, and the extent to which they are utilised, remain unknown. The main work undertaken on Scottish commons concerns grazing, using a pragmatic definition, where such commons were defined as pastures with multiple grazing rights and/or multiple graziers.

There are seven main historic types of common land in Scotland, some of which have similarities to common land in England and Wales.

Commonties

The overwhelming majority of areas of common land in lowland Scotland and the

Highland fringes were *commonties*. A commonty is an area of land where the rights of property or use are shared by two or more neighbouring (though not necessarily adjacent) landowners. They are not therefore truly 'common' land in the sense that anyone can use them, and this distinction meant that it was often very easy for commonties to be divided between landowners after a series of Acts permitting this were passed by the Parliament of Scotland in the 17th century, most notably the 1695 Act for the Division of Commonties. As a result, the number of commonties declined very rapidly in the 18th and 19th centuries.

Common Mosses

Common mosses were areas of bog where the right to dig peat for fuel were shared by neighbouring landowners. They are therefore similar to commonties and most commonties included a common moss. However the difficulties of dividing such wet areas meant that they were left out of many commonty divisions and many common mosses may still survive, un-noticed because of the decline of peat-cutting.

Run Rig

Rig and furrow marks at Buchans Field, Wester Kittochside, an area of Scottish common land

Run rig is a system of agriculture involving the cultivation of adjacent, narrow strips of raised land (rigs). Traditionally adjacent rigs would be used by different farmers and the rigs were periodically re-allocated between them. The system was common throughout Scotland until the 18th century, but survived longer in the Western Highlands, where runrig was often associated with an adjacent area of common hill grazing which was also shared by the same farmers as the runrig.

Scattalds

Scattalds are unique to Shetland and are based on udal law, rather than the feudal law that predominated in the rest of Scotland. However, Scattalds are very similar to commonties and many were divided under the same 1695 Act that allowed for the division of commonties.

Crown Commons

Crown Commons were areas of land held directly by the crown and therefore the common rights that could be used were rights of use rather than rights of property. Unlike commonties, the rights to use crown commons (for example for grazing livestock) were available to anyone, not just the neighbouring landowners. There are no crown commons left in Scotland; those that survived into the 20th century were taken over by the Crown Estate.

Greens and Loans

Greens were small areas of common land near a settlement where livestock could be kept overnight, markets held and other communal activities carried out. Sometimes they were adjacent to drovers' roads near river crossing points or overnight accommodation. Most were genuinely common land with only the Crown holding any title to them. A loan was a common route through private property allowing access to an area of common land or other public place. As the traditional uses of greens and loans declined, they were often absorbed by the neighbouring landowners.

Burgh Commons

Burgh commons were areas of common land where property rights or privileges of use were held by the burgh for their inhabitants. They could include any of the other six types of common land and were sometimes shared with landowners outside the burgh. By the early 19th century, most burgh commons had been appropriated by the wealthy landowners who dominated burgh councils, and very few have survived.

United States

Common land, an English development, was used in many former British colonies, for example in Ireland and the United States. The North American colonies adopted the English laws in establishing their own commons. A famous example is the New Haven Green in New Haven, Connecticut.

View of the Cambridge Common, ca. 1808-9, with Harvard College on the left and Christ Church on the right

Wakefield, MA, town common showing bandstand/gazebo at right and lake at left

Central Burying Ground on the Boston Common in Boston, Massachusetts

Sweden and Finland

A partition unit is a corporation that owns common land. In this case, the land is not state-owned or in joint-ownership under a trust, but is owned by a definite partition unit, a legal partnership whose partners are the participating individual landowners. Common lands and waterways owned by a partition unit were created by an agreement where certain land was reserved for the common use of all adjacent landowners. For the most part, this was due to the Great Partition (Swedish: storskiftet, Finnish: isoja-ko), which started in 1757 and was largely complete by the 1800s. Earlier, the land of a village was divided into narrow stripes of farmland for each to own, with the remainder commonly owned, and work on the land was collective. In the Great Partition, villages were organized as corporations termed partition units (Swedish: skifteslag, Finnish: jakokunta), and land was divided into large chunks that were divided among the households (commoners) for individual cultivation and habitation. Land or waterways that remained undivided was kept by the partition unit as commons, owned by the partition unit. Later, Gustaf III claimed the yet unclaimed forest for the Crown - this was the origin of the large forest holdings of the state in Sweden and Finland. Today, partition units are a common way of owning waterways.

References

- Nguemaleu, Raoul-Abelin Choumin; Montheu, Lionel (2014-05-09). Roadmap to Greener Computing. CRC Press. p. 170. ISBN 9781466506848.

- Holmes, John J. (2008). Reduction of a Ship's Magnetic Field Signatures - Volume 23 of Synthesis lectures on computational electromagnetics. Morgan & Claypool Publishers. p. 19. ISBN 978-1-59829-248-0. Retrieved January 3, 2011.

- Wright, Craig; Kleiman, Dave; Shyaam, Sundhar R.S. (December 2008). "Overwriting Hard Drive Data: The Great Wiping Controversy". Lecture Notes in Computer Science (Springer Berlin / Heidelberg): 243–257. doi:10.1007/978-3-540-89862-7_21. ISBN 978-3-540-89861-0.

- John Millar Carroll (1998). Minimalism Beyond the Nurnberg Funnel. Cambridge, Mass.: MIT Press. ISBN 0-262-03249-X. Retrieved 2007-11-21.

- Visionaries and planners : the garden city movement and the modern community, Stanley Buder. New York: Oxford University Press, 1990. ISBN 0-19-506174-8

- Overton, Mark (1996). Agricultural Revolution in England: The transformation if the agrarian economy 1500-1850. Cambridge University Press. ISBN 978-0-521-56859-3.

- The Electronic Recycling Association is doing computer recycling in Calgary, Edmonton, Winnipeg, Vancouver, retrieved 19 December 2015

- Lohr, Steve (1993-04-14). "Recycling Answer Sought for Computer Junk". The New York Times. ISSN 0362-4331. Retrieved 2015-07-29.

- "Bulgaria opens largest WEEE recycling factory in Eastern Europe". www.ask-eu.com. WtERT Germany GmbH. 12 Jul 2010. Retrieved 2015-07-29.

- Goodman, Peter S. (11 Jan 2012). "Where Gadgets Go To Die: E-Waste Recycler Opens New Plant In Las Vegas". The Huffington Post. Retrieved 2015-07-29.

- "On Re-engineering Discarded Computers, Eliminating e-wastes and Open Source Software" (PDF). American Journal of Computing Research Repository. 1 January 2015. Retrieved 26 May 2015.

- Reed, Michael (February 7, 2008). "I'm Glad That IBM Declined to Release the OS/2 Source". OSNews LLC. Archived from the original on February 21, 2014. Retrieved May 30, 2012.

- "National parks". Department of Communications, Information Technology and the Arts. Australian Government. 31 July 2007. Retrieved 2 November 2014.

Various Materials Recycling Strategies

Some strategies that aid in the recycling of materials are curbside collection, single-stream recycling, materials recovery facility and source reduction. All these strategies aim at segregating garbage into categories of biodegradable and non-biodegradable, increasing awareness about waste generation and reducing volumes of garbage. This chapter provides details about each of these strategies and their importance.

Kerbside Collection

Kerbside collection, or curbside collection, is a service provided to households, typically in urban and suburban areas of removing household waste. It is usually accomplished by personnel using purpose built vehicles to pick up household waste in containers acceptable to or prescribed by the municipality.

Kerbside collection in Canberra, Australia

History

Prior to the 20th century the amount of waste generated by a household was relatively small. Household wastes were often simply thrown out the window, buried in the garden or deposited in outhouses. When human concentrations became more dense, waste collectors, called nightmen or gong farmers were hired to collect the night soil from pail closets, performing their duties only at night (hence the name). Meanwhile, disposing of refuse became a problem wherever cities grew. Often refuse was placed in unusable areas just outside the city, such as wetlands and tidal zones. One example is London, which from Roman times disposed of its refuse outside the London Wall beside the River Thames. Another example is 1830s Manhattan, where thousands of hogs were permitted to roam the streets and eat garbage. A small industry developed as "swill children" collected kitchen refuse to sell for pig feed and the rag and bone man traded goods for bones (used for glue) and rags (essential for paper manufacture prior to the invention of wood pulping). Later, in the late nineteenth century, trash was fed to swine in industrial.

As sanitation engineering came to be practised beginning in the mid-19th century and human waste was conveyed from the home in pipes, the gong farmer was replaced by the municipal trash collector as there remained growing amounts of household refuse, including fly ash from coal, which was burnt for home heating. In Paris, the rag and bone man worked side by side with the municipal bin man, though reluctantly: in 1884, Eugène Poubelle introduced the first integrated kerbside collection and recycling system, requiring residents to separate their waste into perishable items, paper and cloth, and crockery and shells. He also established rules for how private collectors and city workers should cooperate and he developed standard dimensions for refuse containers: his name in France is now synonymous with the garbage can. Under Poubelle, food waste and other organics collected in Paris were transported to nearby Saint Ouen where they were composted. This continued well into the 20th century when plastics began to contaminate the waste stream.

From the late-19th century to the mid-20th century, more or less consistent with the rise of consumables and disposable products municipalities began to pass anti-dumping ordinances and introduce kerbside collection. Residents were required to use a variety of refuse containers to facilitate kerbside collection but the main type was a variation of Poubelle's metal garbage container. It was not until the late 1960s that the green bin bag was introduced by Glad. Later, as waste management practices were introduced with the aim of reducing landfill impacts, a range of container types, mostly made of durable plastic, came to be introduced to facilitate the proper diversion of the waste stream. Such containers include blue boxes, green bins and wheelie bins or MGBs.

Over time, waste collection vehicles gradually increased in size from the hand pushed tip cart or English dust cart, a name by which these vehicles are still referred, to large compactor trucks.

Waste Management and Resource Recovery

Kerbside collection is today often referred to as a strategy of local authorities to collect recyclable items from the consumer. Kerbside collection is considered a low-risk strategy to reduce waste volumes and increase recycling rates. Materials are typically collected in large bins, coloured bags, or small open plastic tubs, specifically designated for content.

Recyclable materials that may be separately collected from municipal waste include:

Biodegradable waste component

- Green waste

- Kitchen waste

Glass for collection in Edinburgh, Scotland.

Recyclable materials, depending on location

- Office paper

- Newsprint

- Paperboard

- Corrugated fiberboard

- Plastics (#1 PET, #2 HDPE natural and colored, #3 PVC narrow-necked containers, #4 LDPE, #5 PP, #6 Polystyrene (however not EXPANDED polystyrene, an example of recyclable polystyrene may be a yoghurt pot) #7 other mixed resin plastics)

- Glass

- Copper

- Aluminum

- Steel and Tinplate

- Co-mingled recyclables- can be sorted by a clean materials recovery facility

Kerbside collection of recyclable resources is aimed to recover purer waste streams with higher market value than by other collection methods. If the household incorrectly separates the recyclable elements, the load may have to be put to landfill if it is deemed to be contaminated.

In Somerville, MA all accepted paper, glass, plastic, and metal recycling is picked up from a single bin

Kerbside collection and household recycling schemes are also being used as tools by local authorities to increase the public's awareness of their waste production.

Kerbside collection is commonly considered to be completely environmentally friendly. This may not necessarily be the case as it leads to an increased number of waste collection vehicles on the road, in themselves contributing to global warming through exhaust emissions until the time of their conversion to clean energy.

New and emerging waste treatment technologies such as mechanical biological treatment may offer an alternative to kerbside collection through automated separation of waste in recycling factories.

Usage by Country

Canada

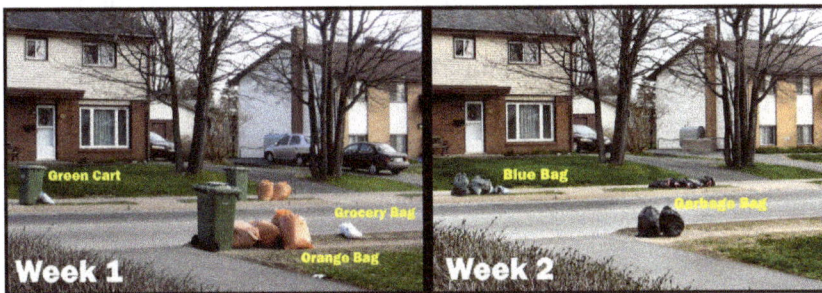

Halifax Regional Municipality (HRM) in Nova Scotia, Canada, with a population of about 375,000, has one of the most complex kerbside collection programmes in North America. Based on the green cart, it requires residents to self-sort refuse and place different types at the kerb on alternating weeks. As shown in the photo at left, week 1 would see the green cart and optional orange bags used for kitchen waste and other organics such as yard waste. Week 2 would permit non-recoverable waste in garbage bags or cans. Blue bags are used for paper, plastic and metal containers. Together with used grocery bags containing newspapers, they may be placed on the kerb either week. In summer, the green cart is emptied weekly due to the prevalence of flies. HRM has achieved a diversion rate of approximately 60 percent by this method.

Calgary, Alberta has adopted "Curbside" Recycling and uses blue bins. The blue cart programme accepts all types of recyclables, including plastics 1-7. It is picked up weekly for the cost of $8.00 per month. This programme is mandatory.

In 1981 Resource Integration Systems (RIS) in collaboration with Laidlaw International tested the first blue box recycling system on 1500 homes in Kitchener, Ontario. Due to the success of the project the City of Kitchener put out a contract for public bid in 1984 for a recycling system city wide. Laidlaw won the bid and continued with the popular blue box recycling system. Today hundreds of cities around the world use the blue box system or a similar variation.

Many Canadian municipalities use "green bins" for kerbside recycling. Others, such as Moncton, use wet/dry waste separation and recovery programmes.

New Zealand

Kerbside collection bins in Dunedin, New Zealand. The yellow-liddied wheelie bin is for non-glass recyclables, and the blue bin is for glass. The two bins are collected on alternating weeks. Official council bags are used for general household waste, and are collected weekly.

In New Zealand, kerbside collection of general refuse and recycling, and in some areas organic waste, is the responsibility of the local city or district council, or private contractors. Practices and collection methods vary widely from council to council and company to company. Some examples of collection are:

- Auckland City Council: Two 240-litre wheelie bins are supplied: a red-lidded bin for general refuse, collected weekly, and a blue-lidded bin for recyclables, collected fortnightly.

- Christchurch City Council: Three wheelie bins are supplied: a 140-litre red-lidded bin for general refuse, a 240-litre yellow-lidded bin for recyclables, and an 80-litre green-lidded bin for organic waste. The organic waste bins are collected weekly, while the recyclables and general refuse bins are collected on alternating weeks.

- Hamilton City Council and Hutt City Council: A 45-litre bin is supplies for recyclables, collected weekly. General refuse is collected weekly using user-pays official council bags.

- Dunedin City Council, Palmerston North City Council and Wellington City Council: Two bins are supplied: a 45-litre or 70-litre bin for glass, and an 80-litre or 240-litre wheelie bin for non-glass recyclables. These two bins are collected on alternating weeks. General refuse is collected weekly using user-pays official council bags.

- Rodney District Council: A 45-litre bin is supplies for recyclables, collected weekly. There is no council collection of general waste, and all general waste collection is carried out by independent companies.

- Taupo District Council: A 45-litre bin is supplies for recyclables, collected weekly. General refuse is collected weekly using user-pays system of orange tags - one orange tag is to be placed on a standard rubbish bag up to 60 litres capacity, or half an orange sticker can be placed on two supermarket bags tied together.

- Upper Hutt City Council: Recycling is to be placed in plastic bags, with paper and cardboard collected in the first week, and plastic, metal and glass in the second week. General refuse is collected weekly using user-pays official council bags.

- Waitakere City Council: A 140-litre wheelie bin is provided for recyclables, collected fortnightly. General refuse is collected weekly using user-pays official council bags.

By 1996 the New Zealand cities of Auckland, Waitakere, North Shore and Lower Hutt had kerbside recycling bins available. In New Plymouth, Wanganui and Upper Hutt recyclable material was collected if placed in suitable bags. By 2007 73% of New Zealanders had access to kerbside recycling.

Kerbside collection of organic waste is carried out by the Mackenzie District Council and the Timaru District Council. Christchurch City Council is introducing the system to their kerbside collection. Other councils are carrying out trials.

United Kingdom

In the United Kingdom, the Household Waste Recycling Act 2003 requires local authorities to provide every household with a separate collection of at least two types of recyclable materials by 2010. There has been criticism in the difference of schemes used in the country such as the colour of bins, whether they are bins boxes or bags, and also the fact that clutter roads and how the additional trucks and collections needed have carbon dioxide emissions too. Some find the colour differences confusing, and people want a national scheme. A typical example is to compare two neighbouring councils in greater Manchester, Bury council and Salford. Bury uses blue for cans, plastic and glass, green for paper and cardboard and brown for garden waste. Salford uses blue for paper and card, brown for cans plastic and glass and pink for garden waste. Most councils use grey or black for general waste, with a few exceptions such as Liverpool, which uses purple for general waste, a colour used by no other council

Another controversial issue in the uk is the frequency of the waste collections. To save money, many councils are cutting the frequency of both general waste and recyclables collections. This has led to problems from larger families, and has led to overflowing and fly tipping. For example, previously, Bury Council collected general waste once a week and recyclables fortnightly. This has now changed to fortnightly for general waste and monthly (every 4 weeks) collection of recyclables.

A few councils are using "forced" recycling, by replacing the large, 240l general waste bin with a smaller 180l or 140l bin, and using the old 240l one for recyclables. This may be made worse by fortnightly collections of the "small" bin, and strict rules such as "No extra bags will be taken" and "Bin lids must be fully closed". Stockport Council is a notable user of this scheme. Their recycling rates have risen substantially as a result, but there are usually complaints from families. Trafford council also use a similar scheme, but the small grey bin is emptied every week. In addition, the two named councils, and more, collect food waste together with garden waste, by sending out kitchen caddies and compostable liners. These prevent food waste (including meat) from going to landfill, and to increase the councils recycling rate. The food and garden waste is usually collected weekly or fortnightly, and is taken to an In Vessel composter or Anaerobic digester, where the waste is turned into soil improver for use on local farms.

In the north west, all the glass collected is used within the UK, around half of the plastics and cans are used in the UK; the rest is sent further afield to Europe or China to be made into new products, and paper and paperboard collected is sent to local paper mills to be made into newspapers, tissues, paperboard and office paper. Again some of the paper will be sent further afield.

Some councils only have 3 bins- general, organic and recyclables. This means that plastics, cans and glass go in the same container as paper and cardboard. Although this is much easier for the residents, there is more sorting required, and the paper quality is sometimes of a low grade due to food contamination or shards of glass in the paper, and so this scheme is criticized.

Also, most councils require residents to remove caps from bottles and rinse them out to avoid smells. This is because the lids are made from a different type of plastic (PP) to the bottle (PET/HDPE) - although by collapsing the bottles and folding them over like toothpaste tubes and rescrewing the caps in place enables the volume of bottles to be drastically reduced, thereby increasing the amount of bottles that can be carried in the recycling bins. In fact many bottlers, especially bottled water companies, have now designed their bottles to be collapsible; though this message has not been effectively disseminated to the consumer. A collapsable bottle takes between 25% and 33% of the space a non-collapsed bottle.

Labels are rarely required to be removed, however. This also means that only plastic bottles are recycled. Councils are still trying to make clear that plastic tubs (yogurts, desserts and spreads), bags and cling film cannot be recycled through the kerbside economically. If too much contamination is collected then this results in the whole vehicle load going to landfill at a high cost. Contamination is usually a problem if recyclables are collected in wheelie bins, as the worker can only look at the top; there may be contamination 'hidden' at the bottom. Councils that use many bags and boxes (Edinburgh) suffer from less contamination but are complicated and the loose paper and cardboard, and recycling bags are blown around, and paper can be wet.

Basque Country

In the province of Gipuzkoa, this system is implanted in many towns as Usurbil, Hernani, Oiartzun, Antzuola, Legorreta, Itsasondo, Zaldibia, Anoeta, Alegia, Irura, Zizurkil, Astigarraga, Ordizia, Oñati and Lezo, where the common used name in basque is atez-atekoa, which means *door-by-door*. Due to the big success in this towns, with more than the 80% of the waste recycled, 34 towns in Gipuzkoa are studying to set this system up in 2013, like Arrasate, Bergara, Aretxabaleta, Eskoriatza, Legazpi, Tolosa or Pasaia.

The atez-ate system consists in hanging each kind of rubbish in a hanger outside the house a certain day or days in a week. For example, in Hernani, they have three days to hang their organic rubbish, two days for plastics and metallics, one for paper and one for rejects residuals.

This system started in the town of Usurbil in the year 2009, due to the incinerator of the region of Gipuzkoa was going to build in this town, exactly in the neighborhood of Zubieta. Three years after, the construction of the incinerator was paralyzed by the government of the region, suggesting that the incinerator was a source of contamination and the high cost of the building.

Single-stream Recycling

Single-stream (also known as "fully commingled" or "single-sort") recycling refers to a system in which all paper fibers, plastics, metals, and other containers are mixed in a collection truck, instead of being sorted by the depositor into separate commodities (newspaper, paperboard, corrugated fiberboard, plastic, glass, etc.) and handled separately throughout the collection process. In single-stream, both the collection and processing systems are designed to handle this fully commingled mixture of recyclables, with materials being separated for reuse at a materials recovery facility (MRF).

Single-stream recycling programs were first developed in several California communities in the 1990s. Subsequently many large and small municipalities across the United States began single-stream programs. As of 2012 there are 248 MRFs operating in the U.S.

Advantages

Proponents of single stream note several advantages:

- Reduced sorting effort by residents may mean more recyclables are placed at the curb and more residents may participate in recycling;

- Reduced collection costs because single-compartment trucks are cheaper to purchase and operate, collection can be automated, and collection routes can be serviced more efficiently;

- Greater fleet flexibility which allows single compartment vehicles to be used to collect recycling, providing greater fleet flexibility and reducing the number of reserve vehicles needed. To avoid confusing customers, a large sign or banner is sometimes used to distinguish when a refuse truck is being used to collect recycling (instead of refuse).

- Worker injuries may decrease because the switch to single stream is often accompanied by a switch from bins to cart-based collection.

- Changing to single stream may provide an opportunity to update the collection and processing system and to add new materials to the list of recyclables accepted; and

- More paper grades may be collected, including junk mail, telephone books and mixed residential paper.

Disadvantages

Potential disadvantages of single stream recycling may include:

- Initial capital cost for:
    - New carts
    - Different collection vehicles
    - Upgrading the processing facility

- Processing costs may increase compared to multiple stream systems

- Possible reduced commodity prices due to contamination of paper or plastic

- Increased downcycling of paper, i.e., use of high quality fibers for low-end uses like boxboard because of presence of contaminants;

- Possible increase in residual rates after processing (chiefly because of increased breakage of glass)

- Potential for diminished public confidence if more recyclables are destined for landfill disposal because of contamination or inability to market materials.

Materials Recovery Facility

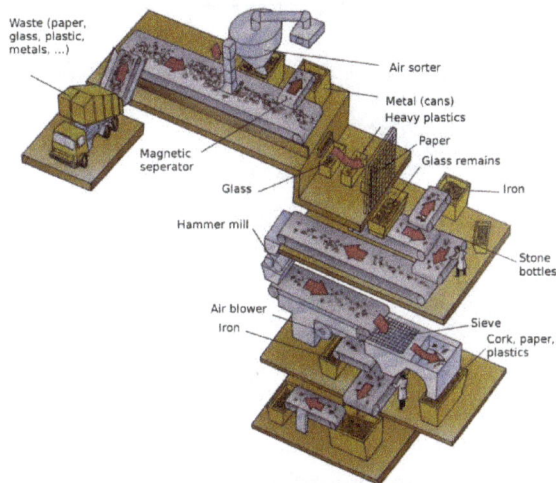

A materials recovery facility for the recycling of domestic waste.

Clean materials recovery facility recycling video.

A materials recovery facility, materials reclamation facility, materials recycling facility or Multi re-use facility (MRF, pronounced "murf") is a specialized plant that receives, separates and prepares recyclable materials for marketing to end-user manufacturers. Generally, there are two different types: clean and dirty materials recovery facilities.

Clean Mrf

A *clean MRF* accepts recyclable commingled materials that have already been separated at the source from municipal solid waste generated by either residential or commercial sources. There are a variety of clean MRFs. The most common are single stream where all recyclable material is mixed, or dual stream MRFs, where source-separated recyclables are delivered in a mixed container stream (typically glass, ferrous metal, aluminum and other non-ferrous metals, PET [No.1] and HDPE [No.2] plastics) and a mixed paper stream including corrugated cardboard boxes, newspapers, magazines, office paper and junk mail. Material is sorted to specifications, then baled, shredded, crushed, compacted, or otherwise prepared for shipment to market.

Mixed-waste Processing Facility (MWPF)

Manual material triage for recycling.

A mixed-waste processing system, sometimes referred to as a dirty MRF, accepts a mixed solid waste stream and then proceeds to separate out designated recyclable materials through a combination of manual and mechanical sorting. The sorted recyclable materials may undergo further processing required to meet technical specifications established by end-markets while the balance of the mixed waste stream is sent to a disposal facility such as a landfill. Today, MWPFs are attracting renewed interest as a way to address low participation rates for source-separated recycling collection systems and prepare fuel products and/or feedstocks for conversion technologies. MWPFs can give communities the opportunity to recycle at much higher rates than has been demonstrated by curbside or other waste collection systems. Advances in technology make today's MWPF different and, in many respects better, than older versions.

The percentage of residuals (unrecoverable recyclable or non-program materials) from a properly operated clean MRF supported by an effective public outreach and education program should not exceed 10% by weight of the total delivered stream and in many cases it can be significantly below 5%. A dirty MRF recovers between 5% and 45% of the incoming material as recyclables, then the remainder is landfilled or otherwise disposed. A dirty MRF can be capable of higher recovery rates than a clean MRF, since it ensures that 100% of the waste stream is subjected to the sorting process, and can target a greater number of materials for recovery than can usually be accommodated by sorting at the source. However, the dirty MRF process results in greater contamination of recyclables, especially of paper. Furthermore, a facility that accepts mixed solid waste is usually more challenging and more expensive to site. Operational costs can be higher because it is more labor-intensive.

Wet Mrf

A wet materials recovery facility

Around 2004, new mechanical biological treatment technologies were beginning to utilise *wet MRFs*. These combine a dirty MRF with water, which acts to densify, separate and clean the output streams. It also hydrocrushes and dissolves biodegradable organics in solution to make them suitable for anaerobic digestion.

History

In the United States, modern MRF's began in the 1970s. Resource Recovery Systems, Inc. operated by Peter Karter was one of "the first materials recovery facilities (MRF) in the US."

Source Reduction

Source reduction is activities designed to reduce the volume, mass, or toxicity of products throughout the life cycle. It includes the design and manufacture, use, and disposal of products with minimum toxic content, minimum volume of material, and/or a longer useful life.

An example of source reduction is use of a Reusable shopping bag at the grocery store; although it uses more material than a single-use disposable bag, the material per use is less.

Synonyms

Pollution Prevention (or P2) and Toxics use reduction are also called source reduction because they address the use of hazardous substances at the source.

Procedures

Source Reduction is achieved through improvements in design, production, use, reuse, recycling, and through Environmentally Preferable Purchasing (EPP). A Life-cycle assessment is useful to help choose among several alternatives and options.

Source Reduction in the United States

In the United States, the Federal Trade Commission offers guidance for labelling claims: "Source reduction" refers to wreducing or lowering the weight, volume or toxicity of a product or package. To avoid being misleading, source reduction claims must qualify the amount of the source reduction and give the basis for any comparison that is made. These principles apply regardless of whether a term like "source reduced" is used.

The Massachusetts Toxics Use Reduction Program (TURA) offers 6 strategies to achieve source reduction:

1. Toxic chemical substitution
2. Production process modification
3. Finished product reformulation
4. Production modernization
5. Improvements in operations and maintenance
6. In-process recycling of production material

References

- Ministry for the Environment (December 2007). Environment New Zealand 2007. Ministry for the Environment (New Zealand). ISBN 978-0-478-30192-2. Retrieved 2008-03-27.
- "Memorandum submitted by Essex Friends of the Earth". www.publications.parliament.uk. Retrieved 19 January 2016.
- Gershman, Brickner & Bratton, Inc., American Chemistry Council, (2015) The Evolution of Mixed Waste Processing Facilities 1970-Today
- de Thomas, Dylan (2013-11-14). "Single Stream in the West." Presentation at Fall 2013 Meeting of Association of Oregon Recyclers, Portland, OR.
- Diehl, Phil (2013-03-05). "Single-stream system increases recycling". San Diego Union-Tribune. San Diego, CA.
- City of Chicago, Illinois. Department of Streets and Sanitation. "What is Single Stream Recycling." Accessed 2013-12-09.
- Montgomery County, Maryland. Division of Solid Waste Services, Rockville, MD (2010). "Comprehensive Solid Waste Management 10 Year Plan: 2009-2019." p. 3-40.

4

Materials Recycling Codes: An Integrated Study

To identify the material from which an item is made and to facilitate faster and easier recycling, recycling codes are added to items. This chapter provides a comprehensive account of all the recycling codes in use, resin identification codes on plastic products and waste hierarchy. This content also helps the reader understand how useful this nomenclature and cataloguing is when recycling products.

Recycling Codes

Recycling codes are used to identify the material from which an item is made, to facilitate easier recycling or other reprocessing. Having a recycling code, the chasing arrows logo or a resin code on an item is not an automatic indicator that a material is recyclable but rather an explanation of what the item is. Such symbols have been defined for batteries, biomatter/organic material, glass, metals, paper, and plastics. Various countries have adopted different codes. For example, the Table below shows the polymer resin codes (plastic) for a country. In the United States there are fewer as ABS is grouped in with other in group 7. Other countries have a more granular recycling code system. For example, China's polymer identification system has seven different classifications of plastic, five different symbols for post-consumer paths, and 140 identification codes The lack of codes in some countries has encouraged those who can fabricate their own plastic products, such as RepRap and other prosumer 3-D printer users, to adopt a voluntary recycling code based on the more comprehensive Chinese system.

Resin Identification Code

The ASTM International Resin Identification Coding System, often abbreviated as the RIC, is a set of symbols appearing on plastic products that identify the plastic resin out of which the product is made. It was developed originally by the Society of the Plastics Industry (now SPI: The Plastics Industry Trade Association) in 1988, but has been administered by ASTM International since 2008.

Sorted household plastic waiting to be hauled away for reprocessing.

Polypropylene lid of a Tic Tac box, with a living hinge and the resin identification code under its flap

History

The Society of the Plastics Industry introduced the Resin Identification Code (RIC) system in 1988 as a growing number of communities were implementing recycling programs. In order to address the concerns of recyclers across the U.S., the RIC system was designed to make it easier for workers in Materials Recovery and Recycling facilities to sort and separate items according to their resin type. Plastics must be recycled separately, with like materials, in order to preserve the material's value and enable its reuse in other products after being recycled.

In its original form, the symbols used as part of the RIC consisted of arrows that cycle clockwise to form a triangle that encloses a number. The number broadly refers to the type of plastic used in the product:

- "1" signifies that the product is made out of polyethylene terephthalate (PET) (beverage bottles, cups, other packaging, etc.)

- "2" signifies high-density polyethylene (HDPE) (bottles, cups, milk jugs, etc.)

- "3" signifies polyvinyl chloride (PVC) (pipes, siding, flooring, etc.)

- "4" signifies low-density polyethylene (LDPE) (plastic bags, six-pack rings, tubing, etc.)

- "5" signifies polypropylene (PP) (auto parts, industrial fibers, food containers, etc.)

- "6" signifies polystyrene (PS) (plastic utensils, Styrofoam, cafeteria trays, etc.)

- "7" signifies other plastics, such as acrylic, nylon, polycarbonate and polylactic acid (PLA).

When a number is omitted, the arrows arranged in a triangle form the universal Recycling Symbol, a generic indicator of recyclability. Subsequent revisions to the RIC have replaced the arrows with a solid triangle, in order to address consumer confusion about the meaning of the RIC, and the belief that the presence of a RIC symbol on an item does not necessarily indicate that it is recyclable.

In 2008, ASTM International took over the administration of the RIC system and eventually issued ASTM D7611 - Standard Practice for Coding Plastic Manufactured Articles for Resin Identification. In 2013 this standard was revised to change the graphic marking symbol of the RIC from the "chasing arrows" of the Recycling Symbol to a solid triangle instead.

Since its introduction, many have used the RIC as a signifier of recyclability, but the presence or absence of a Code on a plastic product does not indicate whether it is recyclable or not.

Table of Resin Codes

Below are the RIC symbols after ASTM's 2013 revision

Resin Identification Number	Resin	Resin Identification Code – Option A	Resin Identification Code – Option B
1	Poly(ethylene terephthalate)	1 PETE	01 PET
2	High density polyethylene	2 HDPE	02 PE-HD
3	Poly(vinyl chloride)	3 V	03 PVC
4	Low density polyethylene	4 LDPE	04 PE-LD
5	Polypropylene	5 PP	05 PP
6	Polystyrene	6 PS	06 PS
7	Other resins	7 OTHER	07 O

Consumer Confusion

In the United States, use of the RIC in the coding of plastics has led to ongoing consumer confusion about which plastic products are recyclable. When many plastics recycling programs were first being implemented in communities across the United States, only plastics with RIC Codes "1" and "2" (polyethylene terephthalate and high-density polyethylene, respectively) were accepted to be recycled. The list of acceptable plastic items has grown since then, and in some areas municipal recycling programs can collect and successfully recycle most plastic products regardless of their RIC Code. This has led some communities to instruct residents to refer to the form of packaging (i.e. "bottles," "tubs," "lids," etc.) when determining what to include in a curbside recycling bin, rather than instructing them to rely on the RIC. To further alleviate consumer confusion, the American Chemistry Council launched the "Recycling Terms & Tools" program to promote standardized language that can be used to educate consumers about how to recycle plastic products.

Possible New Codes

Modifications to the RIC are currently being discussed and developed by ASTM's D20.95 subcommittee on recycled plastics.

In the U.S. the Sustainable Packaging Coalition has also created a "How2Recycle" label in an effort to replace the RIC with that aligns more closely with how the public currently uses the RIC. Rather than indicating what type of plastic resin a product is made out of, the four "How2Recycle" labels indicate whether a plastic product is

- Widely Recycled (meaning greater than 60 percent of the U.S. can recycle the item through a curbside recycling program or municipal drop-off location).

- Limited (meaning only 20-60 percent of the U.S. can recycle the item through a curbside recycling program or municipal drop-off location).

- Not Yet Recycled (meaning less than 20 percent of the U.S. can recycle the item through a curbside recycling program or municipal drop-off location).

- Store Drop-Off (meaning the item can be recycled if brought to participating store drop-off locations, typically at grocery stores).

The "How2Recycle" labels also encourage consumers to check with local facilities to see what plastics each municipal recycling facility can accept.

Waste Hierarchy

The evaluation of processes that protect the environment alongside resource and energy consumption to most favourable to least favourable actions. The hierarchy establishes preferred program priorities based on sustainability. To be sustainable, waste management cannot be solved only with technical end-of-pipe solutions and an integrated approach is necessary.

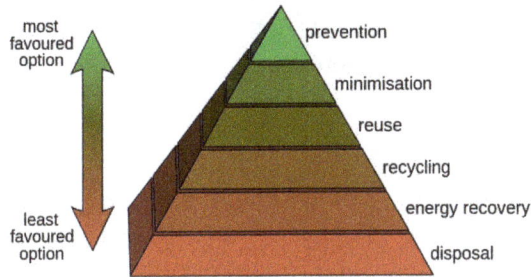

The waste hierarchy.

The three chasing arrows of the international recycling logo. It is sometimes accompanied by the text "reduce, reuse and recycle".t

The waste management hierarchy indicates an order of preference for action to reduce and manage waste, and is usually presented diagrammatically in the form of a pyramid. The hierarchy captures the progression of a material or product through successive stages of waste management, and represents the latter part of the life-cycle for each product.

The aim of the waste hierarchy is to extract the maximum practical benefits from products and to generate the minimum amount of waste. The proper application of the waste hierarchy can have several benefits. It can help prevent emissions of greenhouse gases, reduces pollutants, save energy, conserves resources, create jobs and stimulate the development of green technologies.

Life-cycle Thinking

All products and services have environmental impacts, from the extraction of raw materials for production to manufacture, distribution, use and disposal. Following the waste hierarchy will generally lead to the most resource-efficient and environmentally sound choice but in some cases refining decisions within the hierarchy or departing from it can lead to better environmental outcomes.

Life cycle thinking and assessment can be used to support decision-making in the area of waste management and to identify the best environmental options. It can help policy makers understand the benefits and trade-offs they have to face when making decisions on waste management strategies. Life-cycle assessment provides an approach to ensure that the best outcome for the environment can be identified and put in place. It involves looking at all stages of a product's life to find where improvements can be made to reduce environmental impacts and improve the use or reuse of resources. A key goal is to avoid actions that shift negative impacts from one stage to another. Life cycle thinking can be applied to the five stages of the waste management hierarchy.

For example, life-cycle analysis has shown that it is often better for the environment to replace an old washing machine, despite the waste generated, than to continue to use an older machine which is less energy-efficient. This is because a washing machine's greatest environmental impact is during its use phase. Buying an energy-efficient machine and using low- temperature detergent reduce environmental impacts.

The European Union Waste Framework Directive has introduced the concept of life-cycle thinking into waste policies. This duality approach gives a broader view of all environmental aspects and ensures any action has an overall benefit compared to other options. The actions to deal with waste along the hierarchy should be compatible with other environmental initiatives.

European Union Waste Framework Directive

In 1975, The European Union's Waste Framework Directive (1975/442/EEC) introduced for the first time the waste hierarchy concept into European waste policy. It emphasized the importance of waste minimization, and the protection of the environment and human health, as a priority. Following the 1975 Directive, European Union policy and legislation adapted to the principles of the waste hierarchy.

In 1989, it was formalized into a hierarchy of management options in the European Commission's Community Strategy for Waste Management and this waste strategy was further endorsed in the Commission's review in 1996.

In 2008, the European Union parliament introduced a new five-step waste hierarchy to its waste legislation, Directive 2008/98/EC, which member states must introduce into national waste management laws. Article 4 of the directive lays down a five-step hierarchy of waste management options which must be applied by Member States in this priority order.

Waste prevention, as the preferred option, is followed by reuse, recycling, recovery including energy recovery and as a last option, safe disposal.

Challenges for Local And Regional Authorities

The task of implementing the waste hierarchy in waste management practices within a country may be delegated to the different levels of government (national, regional, local) and to other possible actors including industry, private companies and households. Local and regional authorities can be particularly challenged by the following issues when applying the waste hierarchy approach.

- A coherent waste management strategy must be set up

- Separate collection and sorting systems for many different waste streams need to be established.

- Adequate treatment and disposal facilities must be established.

- An effective horizontal co-operation between local authorities and municipalities and a vertical co-operation between the different levels of government, local to regional and when beneficial, also at the national level need to established

- Finding financing for the establishing or upgrading of expensive sustainable waste management infrastructure to address the needs of managing waste

- A lack of data available on waste management strategies must be overcome and monitoring requirements must be met to implement the waste programs

- The enforcement and control of business plans and practices be established and applied to maximize benefits to the environment and human health

- A lack of administrative capacity at the regional and local level. The lack of finances, information, and technical expertise must be overcome for effective implementation and success of the waste management policies.

Source Reduction

Source reduction involves efforts to reduce hazardous waste and other materials by modifying industrial production. Source reduction methods involve changes in manufacturing technology, raw material inputs, and product formulation. At times, the term "pollution prevention" may refer to source reduction.

Another method of source reduction is to increase incentives for recycling. Many communities in the United States are implementing variable-rate pricing for waste disposal (also known as Pay As You Throw - PAYT) which has been effective in reducing the size of the municipal waste stream.

Source reduction is typically measured by efficiencies and cutbacks in waste. Toxics use reduction is a more controversial approach to source reduction that targets and measures reductions in the upstream use of toxic materials. Toxics use reduction emphasizes the more preventive aspects of source reduction but, due to its emphasis on toxic chemical inputs, has been opposed more vigorously by chemical manufacturers. Toxics use reduction programs have been set up by legislation in some states, e.g., Massachusetts, New

Jersey, and Oregon. The 3 R's represent the 'Waste Hierarchy' which lists the best ways of managing waste from the most to the least desirable. Many of the things we currently throw away could be reused again with just a little thought and imagination.

References

- United Nations Environmental Program (2013). "Guidelines for National Waste Management Strategies Moving from Challenges to Opportunities" (PDF). ISBN 978-92-807-3333-4.

- Department for Environment Food and Rural Affairs. "Guidance on Applying the Waste Hierarchy". gov.uk. Retrieved 2016-05-19.

- "Standard Practice for Coding Plastic Manufactured Articles for Resin Identification". Standard Practice for Coding Plastic Manufactured Articles for Resin Identification. ASTM International. Retrieved 21 January 2016.

- "SPI Resin Identification Code - Guide to Correct Use". SPI: The Plastics Industry Trade Association. SPI: The Plastics Industry Trade Association. Retrieved 21 January 2016.

- Wilhelm, Richard. "Resin Identification Codes - New ASTM Standard Based on Society of the Plastics Industry Code Will Facilitate Recycling". Standardization News (September/October 2008). ASTM International. Retrieved 21 January 2016.

- "Standard Practice for Coding Plastic Manufactured Articles for Resin Identification". Standard Practice for Coding Plastic Manufactured Articles for Resin Identification. ASTM International. Retrieved 14 January 2016.

- Wilson, Allyson. "New "Plastics Recycling Terms & Tools" to Help Communities Recycle More Plastics". American Chemistry Council. American Chemistry Council. Retrieved 14 January 2016.

- "Active Standards under the Jurisdiction of D20.95". Subcommittee D20.95 on Recycled Plastics. ASTM International. Retrieved 14 January 2016.

- "ASTM Plastics Committee Releases Major Revisions to Resin Identification Code (RIC) Standard". ASTM International. ASTM International. Retrieved 21 January 2016.

- "SPI Resin Identification Code - Guide to Correct Use". SPI: The Plastics Industry Trade Association. SPI: The Plastics Industry Trade Association. Retrieved 21 January 2016.

5

Sustainable Recycling of Energy

Recycling is very important in the case of items like batteries as they cannot be degraded and pose an environmental hazard due to the toxic constituents present. This chapter focuses on the topic of battery recycling, rechargeable batteries and energy storage. It provides an in-depth account of the types of batteries and their ease of recycling. There is a section dedicated to renewable energy as well. The topics discussed in the chapter are of great importance to broaden the existing knowledge on recycling.

Battery Recycling

Battery recycling is a recycling activity that aims to reduce the number of batteries being disposed as municipal solid waste. Batteries contain a number of heavy metals and toxichemicals and disposing them by the same process as regular trash has raised concerns over soil contamination and water pollution.

Battery Recycling by Type

Most types of batteries can be recycled. However, some batteries are recycled more readily than others, such as lead–acid automotive batteries (nearly 90% are recycled) and button cells (because of the value and toxicity of their chemicals). Other types, such as alkaline and rechargeable, e.g., nickel–cadmium (Ni-Cd), nickel metal hydride (Ni-MH), lithium-ion (Li-ion) and nickel–zin(Ni-Zn), can also be recycled.

Lead–acid Batteries

These batteries include but are not limited to: car batteries, golf cart batteries, UPS batteries, industrial fork-lift batteries, motorcycle batteries, and commercial batteries. These can be regular lead–acid, sealed lead–acid, gel type, or absorbed glass mat batteries. These are recycled by grinding them, neutralizing the acid, and separating the polymers from the lead. The recovered materials are used in a variety of applications, including new batteries.

Recycling the lead from batteries.

The lead in a lead–acid battery can be recycled. Elemental lead is toxi and should therefore be kept out of the waste stream.

Lead–acid batteries collected by an auto parts retailer for recycling.

Many cities offer battery recycling services for lead–acid batteries. In some jurisdictions, including U.S. states and Canadian provinces, a refundable deposit is paid on batteries. This encourages recycling of old batteries instead of abandonment or disposal with household waste. In the United States, about 99% of lead from used batteries is reclaimed.

Businesses that sell new car batteries may also collect used batteries (or be required to do so by law) for recycling. Some businesses accept old batteries on a "walk-in" basis, as opposed to in exchange for a new battery. Most battery shops and recycling centres pay for scrap batteries. This can be a lucrative business, enticing especially to risk-takers because of the wild fluctuations in the value of scrap lead that can occur overnight. When lead prices go up, scrap batteries become targets for thieves.

Silver Oxide Batteries

Used most frequently in watches, toys, and some medical devices, silver oxide batteries contain a small amount of mercury. Most jurisdictions regulate their handling and disposal to reduce the discharge of mercury into the environment. Silver oxide batteries can be recycled to recover the mercury.

Lithium Ion Batteries

Lithium-ion batteries and lithium iron phosphate (LiFePO4) batteries often contain among other useful metals high-grade copper and aluminium in addition to – depending on the active material – transition metals cobalt and nickel as well as rare earths. To prevent a future shortage of cobalt, nickel, and lithium and to enable a sustainable life cycle of these technologies, recycling processes for lithium batteries are needed. These processes have to regain not only cobalt, nickel, copper, and aluminum from spent battery cells, but also a significant share of lithium. In order to achieve this goal, several unit operations are combined to complex process chains, especially considering the task to recover high rates of valuable materials with regard to involved safety issues.

These unit operations are:

- Deactivation or discharging of the battery (especially in case of batteries from electrivehicles)
- Disassembly of battery systems (especially in case of batteries from electrivehicles)
- Mechanical Processes (including crushing, sorting, and sieving processes)
- Hydrometallurgical processes
- Pyrometallurgical processes

Specifidangers associated with lithium-ion battery recycling processes are: electrical dangers, chemical dangers, burning reactions, and their potential interactions. A complicating factor is the water sensitivity: lithium hexafluorophosphate, a possible electrolyte material, will react with water to form hydrofluoriacid; cells are often immersed in a solvent to prevent this. Once removed, the jelly rolls are separated and the materials removed by ultrasoniagitation, leaving the electrodes ready for melting down and recycling.

Pouch cells are particularly easier to recycle in this way and some people already do this to salvage the copper despite the safety issues.

Battery Composition by type

Type	Fe	Mn	Ni	Zn	Hg	Li	Ag	Cd	Co	Al	Pb	Other	KOH	Paper	Plastic	Alkali	C	Acids	Water	Other
Alkaline	24.8	22.3	0.5	14.9								1.3		1	2.2	5.4	3.7		10.1	14
Zinc–carbon	16.8	15		19.4							0.1	0.8		0.7	4	6	9.2		12.3	15.2
Lithium	50	19	1			2									7		2			19
Mercury-ox-ide	37	1	1	14	31								2		3		1		3	7
Zinc–air	42		1	35	1										4	4	1		10	3
Lithium	60	18	1			3									3		2			13
Alkaline	37	23	1	11	0.6										6	2	2		6	14
Silver oxide	42	2	2	9	0.4		31					4			2	1	0.5		2	4
Nickel-cad-mium	35		22					15				10			10	2			5	11
NiMH	20	1	35	1					4			10			9	4			8	8
Li-ion	22					3			18	5		11					13			28
Lead–acid											65	4			10			16	5	5

Battery Recycling by Location

Country	Return percentage	
	2002	2012
🇨🇭 Switzerland	61 %	73 %
🇧🇪 Belgium	59 %	-
🇸🇪 Sweden	55 %	-
🇩🇪 Germany	39 %	44 %
🇦🇹 Austria	44 %	-
🇳🇱 Netherlands	32 %	-
🇬🇧 United Kingdom	-	32 %
🇫🇷 France	16 %	-
🇨🇦 Canada	3 %	5.6 %

European Union

In 2006 the EU passed the Battery Directive of which one of the aims is a higher rate of battery recycling. The EU directive states that at least 25% of all the EU's used batteries must be collected by 2012, and rising to no less than 45% by 2016, of which, that at least 50% of them must be recycled.

Channel Islands

In early 2009 Guernsey took the initiative by setting up the Longue Hougue recycling facility, which, among other functions, offers a drop-off point for used batteries so they can be recycled off island. The resulting publicity meant that a lot of people complied with the request to dispose of batteries responsibly.

United Kingdom

From April 2005 to March 2008, the UK non-governmental body WRAP conducted trials of battery recycling methods around the UK. The methods tested were: Kerbside, retail drop-off, community drop-off, postal, and hospital and fire station trials. The kerbside trials collected the most battery mass, and were the most well-received and understood by the public. The community drop-off containers which were spread around local community areas were also relatively successful in terms of mass of batteries collected. The lowest performing were the hospital and fire service trials (although these served their purpose very well for specialized battery types like hearing aid and smoke alarm batteries). Retail drop off trials were the second most effective (by volume) method but one of the least well received and used by the public. Both the kerbside and postal trials received the highest awareness and community support.

Household batteries can be recycled in United Kingdom at council recycling sites as well as at some shops and shopping centres—e.g., Dixons, Currys, The Link and PWorld.

A scheme started in 2008 by a large retail company allowed household batteries to be posted free of charge in envelopes available at their shops. This scheme was cancelled at the request of the Royal Mail because of hazardous industrial battery waste being sent as well as household batteries.

An EU directive on batteries that came into force in 2009 means producers must pay for the collection, treatment, and recycling of batteries. This has yet to be ratified into UK law however, so there is currently no real incentive for producers to provide the necessary services.

From 1 February 2010 batteries can be recycled anywhere the Be Positive sign appears. Shops and online retailers that sell more than 32 kilograms of batteries a year must offer facilities to recycle batteries. This is equivalent to one pack of four AA batteries a day. Shops which sell this amount must by law provide recycling facilities as of 1 February 2010.

In Great Britain an increasing number of shops (Argos, Homebase, B&Q, Tesco, and Sainsbury's) are providing battery return boxes and cylinders for their customers.

North America

The rechargeable battery industry has formed the Rechargeable Battery Recycling Corporation (RBRC), which operates a free battery recycling program called Call2Recycle throughout the United States and Canada. RBRprovides businesses with prepaid shipping containers for rechargeable batteries of all types while consumers can drop off batteries at numerous participating collection centers. It claims that no component of any recycled battery eventually reaches a landfill.

A study estimated battery recycling rates in Canada based on RBRdata. In 2002, it wrote, the collection rate was 3.2%. This implies that 3.2% of rechargeable batteries were recycled, and the rest were thrown in the trash. By 2005, it concluded, the collection rate had risen to 5.6%.

In 2009, Kelleher Environmental updated the study. The update estimates the following. "Collection rate values for the 5 [and] 15 year hoarding assumptions respectively are: 8% to 9% for NiCd batteries; 7% to 8% for NiMH batteries; and 45% to 72% for lithium ion and lithium polymer batteries combined. Collection rates through the [RBRC] program for all end of life small sealed lead acid (SLA) consumer batteries were estimated at 10% for 5 year and 15 year hoarding assumptions. [...] It should also be stressed that these figures do not take collection of secondary consumer batteries through other sources into account, and actual collection rates are likely higher than these values."

A November 2011 *New York Times* article reported that batteries collected in the United States are increasingly being transported to Mexico for recycling as a result of a widening gap between the strictness of environmental and labor regulations between the two countries.

In 2015, Energizer announced availability of disposable AAA and AA alkaline batteries made with 3.8% to 4% (by weight) of recycled batteries, branded as EcoAdvanced.

Japan

Japan does not have a single national battery recycling law, so the advice given is to follow local and regional statutes and codes in disposing batteries. The Battery Association of Japan (BAJ) recommends that alkaline, zinc-carbon, and lithium primary batteries can be disposed of as normal household waste. The BAJ's stance on button cell and secondary batteries is toward recycling and increasing national standardisation of procedures for dealing with these types of batteries.

In April 2004 the Japan Portable Rechargeable Battery Recycling Center (JBRC) was created to handle and promote battery recycling throughout Japan. They provide battery recycling containers to shops and other collection points.

Rechargeable Battery

A battery bank used for an uninterruptible power supply in a data center

A rechargeable battery, storage battery, secondary cell, or accumulator is a type of electrical battery which can be charged, discharged into a load, and recharged many times, while a non-rechargeable or primary battery is supplied fully charged, and discarded once discharged. It is composed of one or more electrochemical cells.

The term "accumulator" is used as it accumulates and stores energy through a reversible electrochemical reaction. Rechargeable batteries are produced in many different shapes and sizes, ranging from button cells to megawatt systems connected to stabilize an electrical distribution network. Several different combinations of electrode materials and electrolytes are used, including lead–acid, nickel cadmium (NiCd), nickel metal hydride (NiMH), lithium ion (Li-ion), and lithium ion polymer (Li-ion polymer).

A rechargeable lithium polymer mobile phone battery

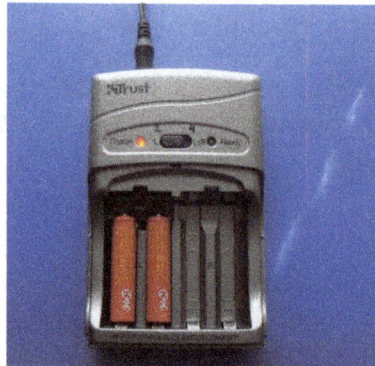

A common consumer battery charger for rechargeable AA and AAA batteries

Rechargeable batteries initially cost more than disposable batteries, but have a much lower total cost of ownership and environmental impact, as they can be recharged inexpensively many times before they need replacing. Some rechargeable battery types are available in the same sizes and voltages as disposable types, and can be used interchangeably with them.

Usage and Applications

Devices which use rechargeable batteries include automobile starters, portable consumer devices, light vehicles (such as motorized wheelchairs, golf carts, electribicycles, and electriforklifts), tools, uninterruptible power supplies, and battery storage power stations. Emerging applications in hybrid internal combustion-battery and electrivehicles drive the technology to reduce cost, weight, and size, and increase lifetime.

Cylindrical cell (18650) prior to assembly. Several thousand of them (lithium ion) form the Tesla Model S battery.

Lithium ion battery monitoring electronics (over- and discharge protection)

Older rechargeable batteries self-discharge relatively rapidly, and require charging before first use; some newer low self-discharge NiMH batteries hold their charge for many months, and are typically sold factory-charged to about 70% of their rated capacity.

Battery storage power stations use rechargeable batteries for load-leveling (storing electrienergy at times of low demand for use during peak periods) and for renewable energy uses (such as storing power generated from photovoltaiarrays during the day to be used at night). Load-leveling reduces the maximum power which a plant must be able to generate, reducing capital cost and the need for peaking power plants.

The US National Electrical Manufacturers Association estimated in 2006 that US demand for rechargeable batteries was growing twice as fast as demand for disposables.

Small rechargeable batteries can power portable electronidevices, power tools, appliances, and so on. Heavy-duty batteries power electrivehicles, ranging from scooters to locomotives and ships. They are used in distributed electricity generation and in standalone power systems.

Charging and Discharging

During charging, the positive active material is oxidized, producing electrons, and the negative material is reduced, consuming electrons. These electrons constitute the current flow in the external circuit. The electrolyte may serve as a simple buffer for internal ion flow between the electrodes, as in lithium-ion and nickel-cadmium cells, or it may be an active participant in the electrochemical reaction, as in lead–acid cells.

A solar-powered charger for rechargeable AA batteries

The energy used to charge rechargeable batteries usually comes from a battery charger using Amains electricity, although some are equipped to use a vehicle's 12-volt Dpower outlet. Regardless, to store energy in a secondary cell, it has to be connected to a Dvoltage source. The negative terminal of the cell has to be connected to the negative terminal of the voltage source and the positive terminal of the voltage source with the positive terminal of the battery. Further, the voltage output of the source must be higher than that of the battery, but not *much* higher: the greater the difference between the power source and the battery's voltage capacity, the faster the charging process, but also the greater the risk of overcharging and damaging the battery.

Chargers take from a few minutes to several hours to charge a battery. Slow "dumb" chargers without voltage or temperature-sensing capabilities will charge at a low rate, typically taking 14 hours or more to reach a full charge. Rapid chargers can typically charge cells in two to five hours, depending on the model, with the fastest taking as little as fifteen minutes. Fast chargers must have multiple ways of detecting when a cell reaches full charge (change in terminal voltage, temperature, etc.) to stop charging before harmful overcharging or overheating occurs. The fastest chargers often incorporate cooling fans to keep the cells from overheating.

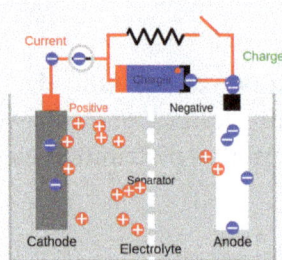

Diagram of the charging of a secondary cell or battery.

Battery charging and discharging rates are often discussed by referencing a "C" rate of current. The rate is that which would theoretically fully charge or discharge the battery in one hour. For example, trickle charging might be performed at C/20 (or a "20 hour" rate), while typical charging and discharging may occur at C/2 (two hours for full capacity). The available capacity of electrochemical cells varies depending on the discharge rate. Some energy is lost in the internal resistance of cell components (plates, electrolyte, interconnections), and the rate of discharge is limited by the speed at which chemicals in the cell can move about. For lead-acid cells, the relationship between time and discharge rate is described by Peukert's law; a lead-acid cell that can no longer sustain a usable terminal voltage at a high current may still have usable capacity, if discharged at a much lower rate. Data sheets for rechargeable cells often list the discharge capacity on 8-hour or 20-hour or other stated time; cells for uninterruptible power supply systems may be rated at 15 minute discharge.

Battery manufacturers' technical notes often refer to voltage per cell (VPC) for the individual cells that make up the battery. For example, to charge a 12 V lead-acid battery (containing 6 cells of 2 V each) at 2.3 VPrequires a voltage of 13.8 V across the battery's terminals.

Non-rechargeable alkaline and zinc–carbon cells output 1.5V when new, but this voltage drops with use. Most NiMH AA and AAA cells are rated at 1.2 V, but have a flatter discharge curve than alkalines and can usually be used in equipment designed to use alkaline batteries.

Damage from Cell Reversal

Subjecting a discharged cell to a current in the direction which tends to discharge it further to the point the positive and negative terminals switch polarity causes a condition called cell reversal. Generally, pushing current through a discharged cell in this way causes undesirable and irreversible chemical reactions to occur, resulting in permanent damage to the cell. Cell reversal can occur under a number of circumstances, the two most common being:

- When a battery or cell is connected to a charging circuit the wrong way around.

- When a battery made of several cells connected in series is deeply discharged.

In the latter case, the problem occurs due to the different cells in a battery having slightly different capacities. When one cell reaches discharge level ahead of the rest, the remaining cells will force the current through the discharged cell.

Many battery-operated devices have a low-voltage cutoff that prevents deep discharges from occurring that might cause cell reversal.

Cell reversal can occur to a weakly charged cell even before it is fully discharged. If the battery drain current is high enough, the cell's internal resistance can create a resistive voltage drop that is greater than the cell's forward emf. This results in the reversal of the cell's polarity while the current is flowing. The higher the required discharge rate of a battery, the better matched the cells should be, both in the type of cell and state of charge, in order to reduce the chances of cell reversal.

In some situations, such as when correcting Ni-Cad batteries that have been previously overcharged, it may be desirable to fully discharge a battery. To avoid damage from the cell reversal effect, it is necessary to access each cell separately: each cell is individually discharged by connecting a load clip across the terminals of each cell, thereby avoiding cell reversal.

Damage During Storage in Fully Discharged State

If a multi-cell battery is fully discharged, it will often be damaged due to the cell reversal effect mentioned above. It is possible however to fully discharge a battery without causing cell reversal—either by discharging each cell separately, or by allowing each cell's internal leakage to dissipate its charge over time.

Even if a cell is brought to a fully discharged state without reversal, however, damage may occur over time simply due to remaining in the discharged state. An example of this is the sulfation that occurs in lead-acid batteries that are left sitting on a shelf for long periods. For this reason it is often recommended to charge a battery that is intended to remain in storage, and to maintain its charge level by periodically recharging it. Since damage may also occur if the battery is overcharged, the optimal level of charge during storage is typically around 30% to 70%.

Depth of Discharge

Depth of discharge (DOD) is normally stated as a percentage of the nominal ampere-hour capacity; 0% DOD means no discharge. Seeing as the usable capacity of a battery system depends on the rate of discharge and the allowable voltage at the end of discharge, the depth of discharge must be qualified to show the way it is to be measured. Due to variations during manufacture and aging, the DOD for complete discharge can change over time or number of charge cycles. Generally a rechargeable battery system will tolerate more charge/discharge cycles if the DOD is lower on each cycle.

Lifespan and Cycle Stability

If batteries are used repeatedly even without mistreatment, they lose capacity as the number of charge cycles increases, until they are eventually considered to have reached the end of their useful life.

Lithium iron phosphate batteries reach according to the manufacturer more than 5000 cycles at respective depth of discharge of 70%. After 7500 cycles with discharge of 85% this still have a spare capacity of at least 80% at a rate of 1 C; which corresponds with a full cycle per day to a lifetime of min. 20.5 years.

The lithium iron phosphate battery Sony Fortelion has after 10,000 cycles at 100% discharge level still a residual capacity of 71%. This battery has been on the market since 2009.

Used in solar batteries Lithium-ion batteries have partly a very high cycle resistance of more than 10,000 charge and discharge cycles and a long service life of up to 20 years.

Plug in America has among drivers of the Tesla Roadster, a survey carried out with respect to the service life of the installed battery. It was found that after 100,000 miles = 160,000 km, the battery still had a remaining capacity of 80 to 85 percent. This was regardless of in which climate zone the car is moved. The Tesla Roadster was built and sold between 2008 and 2012. For its 85-kWh batteries in the Tesla Model S Tesla are 8-year warranty with unlimited mileage.

Varta Storage guarantees its engion battery systems for 14,000 full cycles and a service life of 10 years.

The best-selling electricar is the Nissan Leaf, which is produced since of 2010. Nissan stated in 2015 that until then only 0.01 percent of batteries had to be replaced because of failures or problems and then only because of externally inflicted damage. There are few vehicles that have already covered more than 200,000 km away. These have no problems with the battery.

Recharging Time

BYD e6 taxi. Recharging in 15 Minutes to 80 percent

Electricars like Tesla Model S, Renault Zoe, BMW i3, etc. can recharge their batteries at quick charging stations within 30 minutes to 80 percent.

In laboratories the company StoreDot from Israel reportedly demonstrated the first lab samples of unspecified batteries that can, as of April 2014, be charged in 30 seconds in mobile phones.

Researchers from Singapore in 2014 developed a battery that can be recharged in 2 minutes to 70 percent. The batteries rely on lithium-ion technology. However, the anode and the negative pole in the battery is no longer made of graphite, but a titanium dioxide gel. The gel accelerates the chemical reaction significantly, thus ensuring a faster charging. In particular, these batteries are to be used in electricars. Already in 2012 researchers at the Ludwig-Maximilian-University in Munich have discovered the basiprinciple.

Scientists at Stanford University in California have developed a battery that can be charged within one minute. The anode is made of aluminum and the cathode made of graphite.

The electricar Volar-e of the company Applus + IDIADA, based on the RimaConcept One, contains lithium iron phosphate batteries that can be recharged in 15 minutes.

According to the manufacturer BYD the lithium iron phosphate battery of the electricar e6 is charged at a fast charging station within 15 minutes to 80%, after 40 minutes at 100%.

In 2005, handheld device battery designs by Toshiba were claimed to be able to accept an 80% charge in as little as 60 seconds.

Scientists of university of Oslo from Norway have developed a battery which can be recharged less than one second. According to the scientists this battery would be interesting for example for city buses, which could be loaded at each bus stop, and thus would require only a relatively small battery. A disadvantage is, according to the researchers that the bigger the battery, the greater must be the charging current. Thus, the battery can not be very big. According to the researchers of the new battery could also be used as a buffer in sports car to provide power in the short term. For now, however, the researchers think of applications in small and micro devices.

According to the manufacturer battery of the smartphone OnePlus 3 can be charged from 0 to 60 percent within 30 minutes.

Price History

Lead-acid batteries typically cost €100 / kWh. Li-Ion batteries cost in January 2014, however, typically around 110 € / kWh (150 USD / kWh). The prices for Li-Ion batteries are since 2011 dropped significantly (2011: €500 / kWh, 2012: €350 / kWh, 2013: €200 / kWh) At a conference for electrimobility October 2013 mentioned the trend researcher Lars Thomsen, that Tesla has built its battery at the time 200 USD / kWh (equivalent to €148 / kWh). for the planned for autumn 2016 e-mobile Bolt expects General Motors 145 USD / kWh, and a reduction to 100 USD / kWh by 2022. Reasons for the price decline is the increasing mass production, which has reduced costs through better technologies and economies of scale.

In the German retail LiFePO4 battery cells (as of January 2015) are available from about 420 € / kWh (1.35 € / Ah).

The Powerpack of Tesla costs in spring 2016 250 USD per kWh.

Active Components

The active components in a secondary cell are the chemicals that make up the positive and negative active materials, and the electrolyte. The positive and negative are made up of different materials, with the positive exhibiting a reduction potential and the negative having an oxidation potential. The sum of these potentials is the standard cell potential or voltage.

In primary cells the positive and negative electrodes are known as the cathode and anode, respectively. Although this convention is sometimes carried through to rechargeable systems—especially with lithium-ion cells, because of their origins in primary lithium cells—this practice can lead to confusion. In rechargeable cells the positive electrode is the cathode on discharge and the anode on charge, and vice versa for the negative electrode.

Types

The lead–acid battery, invented in 1859 by French physicist Gaston Planté, is the oldest type of rechargeable battery. Despite having a very low energy-to-weight ratio and a low energy-to-volume ratio, its ability to supply high surge currents means that the cells have a relatively large power-to-weight ratio. These features, along with the low cost, makes it attractive for use in motor vehicles to provide the high current required by automobile starter motors.

The nickel–cadmium battery (NiCd) was invented by Waldemar Jungner of Sweden in 1899. It uses nickel oxide hydroxide and metallicadmium as electrodes. Cadmium is a toxielement, and was banned for most uses by the European Union in 2004. Nickel–cadmium batteries have been almost completely superseded by nickel–metal hydride (NiMH) batteries.

The nickel–metal hydride battery (NiMH) became available in 1989. These are now a common consumer and industrial type. The battery has a hydrogen-absorbing alloy for the negative electrode instead of cadmium.

The lithium-ion battery was introduced in the market in 1991, and it is the choice in most consumer electronics and has the best energy density and a very slow loss of charge when not in use.

Lithium-ion polymer batteries are light in weight, offer slightly higher energy density than Li-ion at slightly higher cost, and can be made in any shape. They are available but have not displaced Li-ion in the market.

Experimental Types

Type	Voltage[a] (V)	Energy density[b] (MJ/kg)	(Wh/kg)	(Wh/L)	Power[c] (W/kg)	E/$[e] (Wh/$)	Self-disch.[f] (%/month)	Cycles[g] (#)	Life[h] (years)
Lithium sulfur	2.0	0.94-1.44	400	350				~1400	
Sodium-ion	3.6			30		3.3		5000+	Testing
Thin film lithium	?		300	959	6000	?p		40000	
Zinc-bromide		0.27-0.31	75-85						
Zinc-cerium	2.5								Under testing
Vanadium redox	1.15-1.55	0.09-0.13	25-35				20%	14,000	10 (stationary)
Sodium-sulfur		0.54	150						
Molten salt	2.58	0.25-1.04	70-290	160	150-220	4.54		3000+	<=20
Silver-zinc	1.86	0.47	130	240					
Quantum Battery (oxide semiconductor)	1.5-3			500	8000(W/L)			100,000	

‡ citations are needed for these parameters

- Nominal cell voltage in V.

- Energy density = energy/weight or energy/size, given in three different units

- Specifipower = power/weight in W/kg

- Energy/consumer price in W·h/US$ (approximately)

- Self-discharge rate in %/month

- Cycle durability in number of cycles

- Time durability in years

- VRLA or recombinant includes gel batteries and absorbed glass mats

- Pilot production

The lithium–sulfur battery was developed by Sion Power in 1994. The company claims superior energy density to other lithium technologies.

The thin film battery (TFB) is a refinement of lithium ion technology by Excellatron. The developers claim a large increase in recharge cycles to around 40,000 and higher charge and discharge rates, at least 5 charge rate. Sustained 60 discharge and 1000peak discharge rate and a significant increase in specifienergy, and energy density.

A smart battery has voltage monitoring circuit built inside. Carbon foam-based lead acid battery: Firefly Energy developed a carbon foam-based lead acid battery with a reported energy density of 30-40% more than their original 38 Wh/kg, with long life and very high power density.

UltraBattery, a hybrid lead-acid battery and ultracapacitor invented by Australia's national science organisation CSIRO, exhibits tens of thousands of partial state of charge cycles and has outperformed traditional lead-acid, lithium and NiMH-based cells when compared in testing in this mode against variability management power profiles. UltraBattery has kW and MW-scale installations in place in Australia, Japan and the U.S.A. It has also been subjected to extensive testing in hybrid electrivehicles and has been shown to last more than 100,000 vehicle miles in on-road commercial testing in a courier vehicle. The technology is claimed to have a lifetime of 7 to 10 times that of conventional lead-acid batteries in high rate partial state-of-charge use, with safety and environmental benefits claimed over competitors like lithium-ion. Its manufacturer suggests an almost 100% recycling rate is already in place for the product.

The potassium-ion battery delivers around a million cycles, due to the extraordinary electrochemical stability of potassium insertion/extraction materials such as Prussian blue.

The sodium-ion battery is meant for stationary storage and competes with lead–acid batteries. It aims at a low total cost of ownership per kWh of storage. This is achieved by a long and stable lifetime. The effective number of cycles is above 5000 and the battery is not damaged by deep discharge. The energy density is rather low, somewhat lower than lead–acid.

The quantum battery (oxide semiconductor) was developed by MJC. It is a small, lightweight cell with a multi-layer film structure and high energy and high power density. It is incombustible, has no electrolyte and generates a low amount of heat during charge.

Its unique feature is its ability to capture electrons physically rather than chemically.

In 2007, Yi Cui and colleagues at Stanford University's Department of Materials Science and Engineering discovered that using silicon nanowires as the anode of a lithium-ion battery increases the anode's volumetricharge density by up to a factor of 10, leading to the development of the nanowire battery.

Another development is the paper-thin flexible self-rechargeable battery combining a thin-film organisolar cell with an extremely thin and highly flexible lithium-polymer battery, which recharges itself when exposed to light.

Ceramatec, a research and development unit of CoorsTek, as of 2009 was testing a battery comprising a chunk of solid sodium metal mated to a sulfur compound by a paper-thin ceramimembrane which conducts ions back and forth to generate a current. The company claimed that it could fit about 40 kilowatt hours of energy into a package about the size of a refrigerator, and operate below 90 °C; and that their battery would allow about 3,650 discharge/recharge cycles (or roughly 1 per day for one decade).

Battery electrodes can be microscopically viewed while bathed in wet electrolytes, resembling conditions inside operating batteries.

In 2014, an Israeli company, StoreDot, claimed to be able to charge batteries in 30 seconds.

Secondary magnesium battery types are an active (2015) topiof research, as a replacement for lithium ion cells.

Aluminium-ion battery types had big success in 2015 in research.

Alternatives

A rechargeable battery is only one of several types of rechargeable energy storage systems. Several alternatives to rechargeable batteries exist or are under development. For uses such as portable radios, rechargeable batteries may be replaced by clockwork mechanisms which are wound up by hand, driving dynamos, although this system may be used to charge a battery rather than to operate the radio directly. Flashlights may be driven by a dynamo directly. For transportation, uninterruptible power supply systems and laboratories, flywheel energy storage systems store energy in a spinning rotor for conversion to electripower when needed; such systems may be used to provide large pulses of power that would otherwise be objectionable on a common electrical grid.

Ultracapacitors—capacitors of extremely high value— are also used; an electriscrewdriver which charges in 90 seconds and will drive about half as many screws as a device using a rechargeable battery was introduced in 2007, and similar flashlights have been produced. In keeping with the concept of ultracapacitors, betavoltaibatteries may be utilized as a method of providing a trickle-charge to a secondary battery, greatly extending the life and energy capacity of the battery system being employed; this type of arrangement is often referred to as a "hybrid betavoltaipower source" by those in the industry.

Ultracapacitors are being developed for transportation, using a large capacitor to store energy instead of the rechargeable battery banks used in hybrid vehicles. One drawback of capacitors compared to batteries is that the terminal voltage drops rapidly; a capacitor that has 25% of its initial energy left in it will have one-half of its initial voltage. By contrast, battery systems tend to have a terminal voltage that does not decline rapidly until nearly exhausted. The undesirable characteristi-complicates the design of power electronics for use with ultracapacitors. However, there are potential benefits in cycle efficiency, lifetime, and weight compared with rechargeable systems. China started using ultracapacitors on two commercial bus routes in 2006; one of them is route 11 in Shanghai.

Flow batteries, used for specialized applications, are recharged by replacing the electrolyte liquid. A flow battery can be considered to be a type of rechargeable fuel cell.

Energy Storage

The Llyn Stwlan dam of the Ffestiniog Pumped Storage Scheme in Wales. The lower power station has four water turbines which can generate a total of 360 MW of electricity for several hours, an example of artificial energy storage and conversion.

Energy storage is the capture of energy produced at one time for use at a later time. A device that stores energy is sometimes called an accumulator. Energy comes in multiple forms including radiation, chemical, gravitational potential, electrical potential, electricity, elevated temperature, latent heat and kinetic. Energy storage involves converting energy from forms that are difficult to store to more conveniently or economically storable forms. Bulk energy storage is dominated by pumped hydro, which accounts for 99% of global energy storage.

Some technologies provide short-term energy storage, while others can endure for much longer.

A wind-up clock stores potential energy (in this case mechanical, in the spring tension), a rechargeable battery stores readily convertible chemical energy to operate a mobile phone, and a hydroelectridam stores energy in a reservoir as gravitational potential energy. Fossil fuels such as coal and gasoline store ancient energy derived from sunlight by organisms that later died, became buried and over time were then converted into these fuels. Food (which is made by the same process as fossil fuels) is a form of energy stored in chemical form.

Ice storage tanks store ice (thermal energy in the form of latent heat) frozen by otherwise wasted energy at night to meet peak daytime demand for cooling. The energy isn't stored directly, but the effect on daytime consumption is equivalent.

History

Prehistory

The energy present at the initial formation of the universe is stored in stars such as the Sun, and is used by humans directly (e.g. through solar heating or sun tanning), or indirectly (e.g. by growing crops, consuming photosynthesized plants or conversion into electricity in solar cells).

As a purposeful activity, energy storage has existed since pre-history, though it was often not explicitly recognized as such. An example of mechanical energy storage is the use of logs or boulders as defensive measures in ancient forts—the logs or boulders were collected at the top of a hill or wall, and the energy thus stored was used to attack invaders who came within range. Certainly, storing dried wood or another source for fire, or preserving edible food or seeds, in dry or cool areas such as in a cave, under rocks or underground, serves as other examples of energy storage.

Recent History

In the twentieth century grid electrical power was largely generated by burning fossil fuel. When less power was required, less fuel was burned. Concerns with air pollution and global warming have spawned the growth of intermittent renewable energy such as solar and wind power. Wind power is uncontrolled and may be generating at a time when no additional power is needed. Interest in storing this power grows as the industry grows.

Off grid electrical use was a niche market in the twentieth century, but in the twenty first century it has expanded. Portable devices are in use all over the world. Solar panels are now a common sight in the rural settings worldwide. Access to electricity is now a question of economics, not location. Powering transportation without burning fuel, however, remains in development.

Methods

Outline

The following list includes natural and other non-commercial types of energy storage. in addition to those designed for use in industry and commerce:

- Mechanical
    - Compressed air energy storage (CAES)
    - Fireless locomotive
    - Flywheel energy storage
    - Gravitational potential energy (device)
    - Hydrauliaccumulator
    - Liquid nitrogen
    - Pumped-storage hydroelectricity
- Electrical
    - Capacitor
    - Superconducting magnetienergy storage (SMES)
- Biological
    - Glycogen
    - Starch
- Electrochemical
    - Flow battery
    - Rechargeable battery
    - Supercapacitor
    - UltraBattery
- Thermal
    - Brick storage heater
    - Cryogeniliquid air or nitrogen

- o   Eutectisystem

- o   Ice Storage

- o   Molten salt

- o   Phase Change Material

- o   Seasonal thermal energy storage

- o   Solar pond

- o   Steam accumulator

- o   Thermal energy storage (general)

- Chemical

  - o   Biofuels

  - o   Hydrated salts

  - o   Hydrogen

  - o   Hydrogen peroxide

  - o   Power to gas

  - o   Vanadium pentoxide

Mechanical Storage

Energy can be stored in water pumped to a higher elevation using pumped storage methods and also by moving solid matter to higher locations. Other commercial mechanical methods include compressing air and flywheels that convert electrienergy into kinetienergy and then back again when electrical demand peaks.

Hydroelectricity

Hydroelectridams with reservoirs can be operated to provide peak generation at times of peak demand. Water is stored in the reservoir during periods of low demand and released when demand is high. The net effect is similar to pumped storage, but without the pumping loss.

While a hydroelectridam does not directly store energy from other generating units, it behaves equivalently by lowering output in periods of excess electricity from other sources. In this mode, dams are one of the most efficient forms of energy storage, because only the timing of its generation changes. Hydroelectriturbines have a start-up time on the order of a few minutes.

Pumped-storage

The Sir Adam Beck Generating Complex at Niagara Falls, Canada, which includes a large pumped storage hydroelectricity reservoir to provide an extra 174 MW of electricity during periods of peak demand.

Worldwide, pumped-storage hydroelectricity (PSH) is the largest-capacity form of active grid energy storage available, and, as of March 2012, the ElectriPower Research Institute (EPRI) reports that PSH accounts for more than 99% of bulk storage capacity worldwide, representing around 127,000 MW. PSH reported energy efficiency varies in practice between 70% and 80%, with claims of up to 87%.

At times of low electrical demand, excess generation capacity is used to pump water from a lower source into a higher reservoir. When demand grows, water is released back into a lower reservoir (or waterway or body of water) through a turbine, generating electricity. Reversible turbine-generator assemblies act as both a pump and turbine (usually a Francis turbine design). Nearly all facilities use the height difference between two water bodies. Pure pumped-storage plants shift the water between reservoirs, while the "pump-back" approach is a combination of pumped storage and conventional hydroelectriplants that use natural stream-flow.

Compressed Air

Devices which use rechargeable batteries include automobile starters, portable consumer devices, light vehicles (such as motorized wheelchairs, golf carts, electribicycles, and electriforklifts), tools, uninterruptible power supplies, and battery storage power stations. Emerging applications in hybrid internal combustion-battery and electrivehicles drive the technology to reduce cost, weight, and size, and increase lifetime.

Compressed air energy storage (CAES) uses surplus energy to compress air for subsequent electricity generation. Small scale systems have long been used in such applications as propulsion of mine locomotives. The compressed air is stored in an underground reservoir.

Compression of air creates heat; the air is warmer after compression. Expansion requires heat. If no extra heat is added, the air will be much colder after expansion. If the heat generated during compression can be stored and used during expansion, efficiency improves considerably. A CAES system can deal with the heat in three ways. Air storage can be adiabatic, diabatic, or isothermal. Another approach uses compressed air to power vehicles.

A compressed air locomotive used inside a mine between 1928 and 1961.*e*

Flywheel Energy Storage

The main components of a typical flywheel.

Flywheel energy storage (FES) works by accelerating a rotor (flywheel) to a very high speed, holding energy as rotational energy. When energy is extracted, the flywheel's rotational speed declines as a consequence of conservation of energy; adding energy correspondingly results in an increase in the speed of the flywheel.

Most FES systems use electricity to accelerate and decelerate the flywheel, but devices that directly use mechanical energy are under consideration.

FES systems have rotors made of high strength carbon-fiber composites, suspended by magnetibearings and spinning at speeds from 20,000 to over 50,000 rpm in a vacuum enclosure. Such flywheels can reach maximum speed ("charge") in a matter of minutes. The flywheel system is connected to a combination electrimotor/generator.

FES systems have relatively long lifetimes (lasting decades with little or no maintenance; full-cycle lifetimes quoted for flywheels range from in excess of 10^5, up to 10^7, cycles of use), high energy density (100–130 W·h/kg, or 360–500 kJ/kg) and power density.

A Flybrid KinetiEnergy Recovery System flywheel. Built for use on Formula 1 racing cars, it is employed to recover and reuse kinetienergy captured during braking.

Gravitational Potential Energy Storage with Solid Masses

Changing the altitude of solid masses can store or release energy via an elevating system driven by an electrimotor/generator.

Companies such as Energy Cache and Advanced Rail Energy Storage (ARES) are working on this. ARES uses rails to move concrete weights up and down. Stratosolar proposes to use winches supported by buoyant platforms at an altitude of 20 kilometers, to raise./lower solid masses. Sink Float Solutions proposes to use winches supported by an ocean barge for taking advantage of a 4 km (13,000 ft) elevation difference between the surface and the seabed. ARES estimated a capital cost for the storage capacity of around 60% of pump storage hydroelectricity, Stratosolar $100/kWh and Sink Float Solutions $25/kWh (4000 m depth) and $50/kWh (with 2000 m depth).

Potential energy storage or gravity energy storage was under active development in 2013 in association with the California Independent System Operator. It examined the movement of earth-filled hopper rail cars driven by electrilocomotives) from lower to higher elevations.

ARES claimed advantages including indefinite storage with no energy losses, low costs when earth/rocks are used and conservation of water resources.

Thermal Storage

Thermal storage is the temporary storage or removal of heat. TES is practical because of water's large heat of fusion: the melting of one metriton of ice (approximately one cubimetre in size) can capture 334 megajoules [MJ] (317,000 BTU) of thermal energy.

An example is Alberta, Canada's Drake Landing Solar Community, for which 97% of the year-round heat is provided by solar-thermal collectors on the garage roofs, with a borehole thermal energy store (BTES) being the enabling technology. STES projects often have paybacks in the four-to-six year range. In Braestrup, Denmark, the community's solar district heating system also utilizes STES, at a storage temperature of 65 °(149 °F). A heat pump, which is run only when there is surplus wind power available on the national grid, is used to raise the temperature to 80 °(176 °F) for distribution. When surplus wind generated electricity is not available, a gas-fired boiler is used. Twenty percent of Braestrup's heat is solar.

District heating accumulation tower from Theiss near Krems an der Donau in Lower Austria with a thermal capacity of 2 GWh

Latent Heat Thermal Energy Storage (LHTES)

Latent heat thermal energy storage systems works with materials with high latent heat (heat of fusion) capacity, known as phase change materials (PCMs). The main advantage of these materials is that their latent heat storage capacity is much more than sensible heat. In a specifitemperature range, phase changes from solid to liquid absorbs a large amount of thermal energy for later use.

Electrochemical

Rechargeable battery

A rechargeable battery bank used as an uninterruptible power supply in a data center

A rechargeable battery, comprises one or more electrochemical cells. It is known as a 'secondary cell' because its electrochemical reactions are electrically reversible. Rechargeable batteries come in many different shapes and sizes, ranging from button cells to megawatt grid systems.

Rechargeable batteries have lower total cost of use and environmental impact than non-rechargeable (disposable) batteries. Some rechargeable battery types are available in the same form factors as disposables. Rechargeable batteries have higher initial cost but can be recharged very cheaply and used many times.

Common rechargeable battery chemistries include:

- Lead–acid battery: Lead acid batteries hold the largest market share of electris-torage products. A single cell produces about 2V when charged. In the charged state the metallilead negative electrode and the lead sulfate positive electrode are immersed in a dilute sulfuriacid (H2SO4) electrolyte. In the discharge process electrons are pushed out of the cell as lead sulfate is formed at the negative electrode while the electrolyte is reduced to water.

- Nickel–cadmium battery (NiCd): Uses nickel oxide hydroxide and metallicadmium as electrodes. Cadmium is a toxielement, and was banned for most uses by the European Union in 2004. Nickel–cadmium batteries have been almost completely replaced by nickel–metal hydride (NiMH) batteries.

- Nickel–metal hydride battery (NiMH): First commercial types were available in 1989. These are now a common consumer and industrial type. The battery has a hydrogen-absorbing alloy for the negative electrode instead of cadmium.

- Lithium-ion battery: The choice in many consumer electronics and have one of the best energy-to-mass ratios and a very slow self-discharge when not in use.

- Lithium-ion polymer battery: These batteries are light in weight and can be made in any shape desired.

Flow Battery

A flow battery operates by passing a solution over a membrane where ions are exchanged to charge/discharge the cell. Cell voltage is chemically determined by the Nernst equation and ranges, in practical applications, from 1.0 to 2.2 V. Its storage capacity is a function of the volume of the tanks holding the solution.

A flow battery is technically akin both to a fuel cell and an electrochemical accumulator cell. Commercial applications are for long half-cycle storage such as backup grid power.

Supercapacitor

Supercapacitors, also called electridouble-layer capacitors (EDLC) or ultracapacitors, are generiterms for a family of electrochemical capacitors that do not have conventional solid dielectrics. Capacitance is determined by two storage principles, double-layer capacitance and pseudocapacitance.

Supercapacitors bridge the gap between conventional capacitors and rechargeable batteries. They store the most energy per unit volume or mass (energy density) among capacitors. They support up to 10,000 farads/1.2 volt, up to 10,000 times that of electro-lyticapacitors, but deliver or accept less than half as much power per unit time (power density).

One of a fleet of electricapabuses powered by supercapacitors, at a quick-charge station-bus stop, in service during Expo 2010 Shanghai China. Charging rails can be seen suspended over the bus.

While supercapacitors have energy densities that are approximately 10% of batteries, their power density is generally 10 to 100 times greater. This results in much shorter charge/discharge cycles. Additionally, they will tolerate many more charge and discharge cycles than batteries.

Supercapacitors support a broad spectrum of applications, including:

- Low supply current for memory backup in statirandom-access memory (SRAM)

- Power for cars, buses, trains, cranes and elevators, including energy recovery from braking, short-term energy storage and burst-mode power delivery

UltraBattery

The UltraBattery is a hybrid lead-acid cell and carbon-based ultracapacitor (or super-capacitor) invented by Australia's national research body, the Commonwealth Scienti-fiand Industrial Research Organisation (CSIRO). The lead-acid cell and ultracapacitor share the sulfuriacid electrolyte and both are packaged into the same physical unit. The UltraBattery can be manufactured with similar physical and electrical characteristics to conventional lead-acid batteries making it possible to cost-effectively replace many lead-acid applications.

The UltraBattery tolerates high charge and discharge levels and endures large numbers of cycles, outperforming previous lead-acid cells by more than an order of magnitude. In hybrid-electrivehicle tests, millions of cycles have been achieved. The UltraBattery is also highly tolerant to the effects of sulfation compared with traditional lead-acid cells. This means it can operate continuously in partial state of charge whereas traditional lead-acid batteries are generally held at full charge between discharge events. It is generally electrically inefficient to fully charge a lead-acid battery so by decreasing time spent in the top region of charge the UltraBattery achieves high efficiencies, typically between 85 and 95% DC-DC.

The UltraBattery can work across a wide range of applications. The constant cycling and fast charging and discharging necessary for applications such as grid regulation and leveling and electrivehicles can damage chemical batteries, but are well handled by the ultracapacitive qualities of UltraBattery technology. The technology has been installed in Australia and the US on the megawatt scale, performing frequency regulation and renewable smoothing applications.

Other Chemical

Power to Gas

Power to gas is a technology which converts electricity into a gaseous fuel such as hydrogen or methane. The three commercial methods use electricity to reduce water into hydrogen and oxygen by means of electrolysis.

In the first method, hydrogen is injected into the natural gas grid or is used in transport or industry. The second method is to combine the hydrogen with carbon dioxide to produce methane using a methanation reaction such as the Sabatier reaction, or biological methanation, resulting in an extra energy conversion loss of 8%. The methane may then be fed into the natural gas grid. The third method uses the output gas of a wood gas generator or a biogas plant, after the biogas upgrader is mixed with the hydrogen from the electrolyzer, to upgrade the quality of the biogas.

Hydrogen

Atomihydrogen is a form of stored energy. It can produce electricity via a hydrogen fuel cell.

At penetrations below 20% of the grid demand, renewables do not severely change the economics; but beyond about 20% of the total demand, external storage becomes important. If these sources are used to make ionihydrogen, they can be freely expanded. A 5-year community-based pilot program using wind turbines and hydrogen generators began in 2007 in the remote community of Ramea, Newfoundland and Labrador. A similar project began in 2004 on Utsira, a small Norwegian island.

Energy losses involved in the hydrogen storage cycle come from the electrolysis of water, liquification or compression of the hydrogen and conversion to electricity.

About 50 kW·h (180 MJ) of solar energy is required to produce a kilogram of hydrogen, so the cost of the electricity is crucial. At $0.03/kWh, a common off-peak high-voltage line rate in the United States, hydrogen costs $1.50 a kilogram for the electricity, equivalent to $1.50/gallon for gasoline. Other costs include the electrolyzer plant, hydrogen compressors or liquefaction, storage and transportation.

Underground hydrogen storage is the practice of hydrogen storage in underground caverns, salt domes and depleted oil and gas fields. Large quantities of gaseous hydrogen have been stored in underground caverns by Imperial Chemical Industries for many years without any difficulties. The European Hyunder project indicated in 2013 that storage of wind and solar energy using underground hydrogen would require 85 caverns. This amount of storage is well beyond what pumped-storage hydroelectricity and compressed air energy storage systems can provide.

Methane

Methane is the simplest hydrocarbon with the molecular formula CH_4. Methane is more easily stored than hydrogen and the transportation. Storage and combustion infrastructure (pipelines, gasometers, power plants) are mature.

Synthetinatural gas (syngas or SNG) can be created in a multi-step process, starting with hydrogen and oxygen. Hydrogen is then reacted with carbon dioxide in a Sabatier process, producing methane and water. Methane can be stored and later used to produce electricity. The resulting water is recycled, reducing the need for water. In the electrolysis stage oxygen is stored for methane combustion in a pure oxygen environment at an adjacent power plant, eliminating nitrogen oxides.

Methane combustion produces carbon dioxide (CO_2) and water. The carbon dioxide can be recycled to boost the Sabatier process and water can be recycled for further electrolysis. Methane production, storage and combustion recycles the reaction products.

The CO_2 has economivalue as a component of an energy storage vector, not a cost as in carbon capture and storage.

Power to Liquid

Power to liquid is similar to power to gas, however the hydrogen produced by electrolysis from wind and solar electricity isn't converted into gases such as methane but into liquids such as methanol. Methanol is easier in handling than gases and requires less safety precautions than hydrogen. It can be used for transportation, including aircrafts, but also for industrial purposes or in the power sector.

Biofuels

Various biofuels such as biodiesel, vegetable oil, alcohol fuels, or biomass can replace fossil fuels. Various chemical processes can convert the carbon and hydrogen in coal, natural gas, plant and animal biomass and organiwastes into short hydrocarbons suitable as replacements for existing hydrocarbon fuels. Examples are Fischer–Tropsch diesel, methanol, dimethyl ether and syngas. This diesel source was used extensively in World War II in Germany, which faced limited access to crude oil supplies. South Africa produces most of the country's diesel from coal for similar reasons. A long term oil price above US$35/bbl may make such large scale synthetiliquid fuels economical.

Aluminium, Boron, Silicon, and Zinc

Aluminium, Boron, silicon, lithium, and zinhave been proposed as energy storage solutions.

Electrical Methods

Capacitor

This mylar-film, oil-filled capacitor has very low inductance and low resistance, to provide the high-power (70 megawatts) and the very high speed (1.2 microsecond) discharges needed to operate a dye laser.

A capacitor (originally known as a 'condenser') is a passive two-terminal electrical component used to store energy electrostatically. Practical capacitors vary widely, but all contain at least two electrical conductors (plates) separated by a dielectri(i.e., insulator). A capacitor can store electrienergy when disconnected from its charging circuit, so it can be used like a temporary battery, or like other types of rechargeable energy storage system. Capacitors are commonly used in electronidevices to maintain power supply while batteries change. (This prevents loss of information in volatile memory.) Conventional capacitors provide less than 360 joules per kilogram, while a conventional alkaline battery has a density of 590 kJ/kg.

Capacitors store energy in an electrostatifield between their plates. Given a potential difference across the conductors (e.g., when a capacitor is attached across a battery), an electrifield develops across the dielectric, causing positive charge (+Q) to collect on one plate and negative charge (-Q) to collect on the other plate. If a battery is attached to a capacitor for a sufficient amount of time, no current can flow through the capacitor. However, if an accelerating or alternating voltage is applied across the leads of the capacitor, a displacement current can flow.

Capacitance is greater given a narrower separation between conductors and when the conductors have a larger surface area. In practice, the dielectribetween the plates emits a small amount of leakage current and has an electrifield strength limit, known as the breakdown voltage. The conductors and leads introduce undesired inductance and resistance.

Research is assessing the quantum effects of nanoscale capacitors for digital quantum batteries.

Superconducting Magnetics

Superconducting magnetienergy storage (SMES) systems store energy in a magneti-field created by the flow of direct current in a superconducting coil that has been cooled to a temperature below its superconducting critical temperature. A typical SMES system includes a superconducting coil, power conditioning system and refrigerator. Once the superconducting coil is charged, the current does not decay and the magnetienergy can be stored indefinitely.

The stored energy can be released to the network by discharging the coil. The associated inverter/rectifier accounts for about 2–3% energy loss in each direction. SMES loses the least amount of electricity in the energy storage process compared to other methods of storing energy. SMES systems offer round-trip efficiency greater than 95%.

Due to the energy requirements of refrigeration and the cost of superconducting wire, SMES is used for short duration storage such as improving power quality. It also has applications in grid balancing.

Interseasonal Thermal Storage

Seasonal thermal energy storage (STES) allows heat or cold to be used months after it was collected from waste energy or natural sources. The material can be stored in contained aquifers, clusters of boreholes in geological substrates such as sand or crystalline bedrock, in lined pits filled with gravel and water, or water-filled mines.

Applications

Mills

A more recent application is the control of waterways to drive water mills for processing grain or powering machinery. Complex systems of reservoirs and dams were constructed to store and release water (and the potential energy it contained) when required.

Home Energy Storage

Home energy storage is expected to become increasingly present given the growing importance of distributed generation (especially photovoltaics) and the important share of energy consumption in buildings . A household equipped with photovoltaics can achieve a maximum electricity self-sufficiency of about 40%. To reach higher levels of self-sufficiency,energy storage is needed, given the mismatch between energy consumption and energy production from photovoltaics . In 2015, multiple manufacturers announced rechargeable battery systems for storing energy, generally to hold surplus energy from home solar/wind generation.

Tesla Motors announced the first two models of theTesla Powerwall. One is a 10 kWh weekly cycle version for backup applications and the other is a 7 kWh version for daily cycle applications.

Enphase Energy announced an integrated system that allows home users to store, monitor and manage electricity. The system stores 1.2 kWh hours of energy and 275W/500W power output.

Grid electricity

Renewable Energy Storage

Construction of the Salt Tanks which provide efficient thermal energy storage so that output can be provided after the sun goes down, and output can be scheduled to meet demand requirements. The 280 MW Solana Generating Station is designed to provide six hours of energy storage. This allows the plant to generate about 38 percent of its rated capacity over the course of a year.

The 150 MW Andasol solar power station is a commercial parabolitrough solar thermal power plant, located in Spain. The Andasol plant uses tanks of molten salt to store solar energy so that it can continue generating electricity even when the sun isn't shining.

The largest source and the greatest store of renewable energy is provided by hydroelectridams. A large reservoir behind a dam can store enough water to average the annual flow of a river between dry and wet seasons. A very large reservoir can store enough water to average the flow of a river between dry and wet years. While a hydroelectridam does not directly store energy from intermittent sources, it does balance the grid by lowering its output and retaining its water when power is generated by solar or wind. If wind or solar generation exceeds the regions hydroelectricapacity, then some additional form of energy balancing will be needed.

Many renewable energy sources (notably solar and wind) produce variable power. Storage systems can level out the imbalances between supply and demand that this causes. Electricity must be used as it is generated or converted immediately into storable forms.

The first method of large scale storage was pumped-storage hydroelectricity. Areas of the world such as Norway, Washington and Oregon states and Wales have used geographifeatures to supply elevated reservoirs, using electricity to fill them. When needed, the water passes through generators and converts the potential energy back to electricity. Pumped storage in Norway has an instantaneous capacity of 25–30 GW expandable to 60 GW—enough to be "Europe's battery".

Common forms of renewable energy storage include pumped-storage hydroelectridams, rechargeable batteries, thermal storage including molten salts which can efficiently store and release very large quantities of heat energy, and compressed air energy storage, flywheels, cryogenisystems and superconducting magneticoils.

Surplus power can also be converted into methane (sabatier process) with stockage in the natural gas network.

In 2011, the Bonneville Power Administration in Northwestern United States created an experimental program to absorb excess wind and hydro power generated at night or during stormy periods that are accompanied by high winds. Under central control, home appliances absorb surplus energy by heating ceramibricks in special space heaters to hundreds of degrees and by boosting the temperature of modified hot water heater tanks. After charging, the appliances provide home heating and hot water as needed. The experimental system was created as a result of a severe 2010 storm that overproduced renewable energy to the extent that all conventional power sources were shut down, or in the case of a nuclear power plant, reduced to its lowest possible operating level, leaving a large area running almost completely on renewable energy.

Another advanced method used at the Solar Project in the United States and the Solar Tres Power Tower in Spain uses molten salt to store thermal energy captured from the sun and then convert it and dispatch it as electrical power. The system pumps molten salt through a tower or other special conduits to be heated by the sun. Insulated tanks store the solution. Electricity is produced by turning water to steam that is fed to turbines.

Since the early 21st century batteries have been applied to utility scale load-leveling and frequency regulation capabilities.

In vehicle-to-grid storage, electrivehicles that are plugged into the energy grid can deliver stored electrical energy from their batteries into the grid when needed.

Generation

Chemical fuels remain the dominant form of energy storage for electricity generation. Natural gas is crowding out other forms, such as oil and coal.

Air Conditioning

thermal energy storage (TES) can be used for air conditioning. It is most widely used for cooling single large buildings and/or groups of smaller buildings. Commercial air conditioning systems are the biggest contributors to peak electrical loads. In 2009, thermal storage was used in over 3,300 buildings in over 35 countries. It works by creating ice at night and using the ice to for cooling during the hotter daytime periods.

The most popular technique is ice storage, which requires less space than water and is less costly than fuel cells or flywheels. In this application, a standard chiller runs at night to produce an ice pile. Water then circulates through the pile during the day to chill water that would normally be the chiller's daytime output.

A partial storage system minimizes capital investment by running the chillers nearly 24 hours a day. At night, they produce ice for storage and during the day they chill water. Water circulating through the melting ice augments the production of chilled water. Such a system makes ice for 16 to 18 hours a day and melts ice for six hours a day. Capital expenditures are reduced because the chillers can be just 40 - 50% of the size needed for a conventional, no-storage design. Storage sufficient to store half a day's available heat is usually adequate.

A full storage system shuts off the chillers during peak load hours. Capital costs are higher, as such a system requires larger chillers and a larger ice storage system.

This ice is produced when electrical utility rates are lower. Off-peak cooling systems can lower energy costs. The U.S. Green Building Council has developed the Leadership in Energy and Environmental Design (LEED) program to encourage the design of reduced-environmental impact buildings. Off-peak cooling may help toward LEED Certification.

Thermal storage for heating is less common than for cooling. An example of thermal storage is storing solar heat to be used for heating at night.

Latent heat can also be stored in technical phase change materials (PCMs). These can be encapsulated in wall and ceiling panels, to moderate room temperatures.

Transport

Liquid hydrocarbon fuels are the most commonly used forms of energy storage for use in transportation. Other energy carriers such as hydrogen can be used to avoid producing greenhouse gases.

Electronics

Capacitors are widely used in electronicircuits for blocking direct current while allowing alternating current to pass. In analog filter networks, they smooth the output of power supplies. In resonant circuits they tune radios to particular frequencies. In electripower transmission systems they stabilize voltage and power flow.

Use cases

The United States Department of Energy International Energy Storage Database (IESDB), is a free-access database of energy storage projects and policies funded by the United States Department of Energy Office of Electricity and Sandia National Labs.

Metrics

A metrifor calculating the energy efficiency of storage systems is Energy Storage On Energy Invested (ESOI) which is the useful energy used to make the storage system divided into the lifetime energy storage. For lithium ion batteries this is around 10, and for lead acid batteries it is about 2. Other forms of storage such as pumped hydroelectristorage generally have higher ESOI, such as 210.

The economivaluation of large-scale applications (including pumped hydro storage and compressed air) considers benefits including: wind curtailment avoidance, grid congestion avoidance, price arbitrage and carbon free energy delivery. In one technical assessment by the Carnegie Mellon Electricity Industry Centre, economigoals could be met with batteries if energy storage were achievable at a capital cost of $30 to $50 per kilowatt-hour of storage capacity.

Research

Germany

The German Federal government has allocated €200M (approximately US$270M) for advanced research, as well as providing a further €50M to subsidize battery storage for use with residential rooftop solar panels, according to a representative of the German Energy Storage Association.

Siemens AG commissioned a production-research plant to open in 2015 at the *Zentrum für Sonnenenergie und Wasserstoff (ZSW,* the German Center for Solar Energy and Hydrogen Research in the State of Baden-Württemberg), a university/industry collaboration in Stuttgart, Ulm and Widderstall, staffed by approximately 350 scientists, researchers, engineers, and technicians. The plant develops new near-production manufacturing materials and processes (NPMM&P) using a computerized Supervisory Control and Data Acquisition (SCADA) system. Its goals will enable the expansion of rechargeable battery production with both increased quality and reduced manufacturing costs.

United States

In 2014, research and test centers opened to evaluate energy storage technologies. Among them was the Advanced Systems Test Laboratory at the University of Wisconsin at Madison in Wisconsin State, which partnered with battery manufacturer Johnson Controls. The laboratory was created as part of the university's newly opened Wisconsin Energy Institute. Their goals include the evaluation of state-of-the-art and next generation electrivehicle batteries, including their use as grid supplements.

The State of New York unveiled its New York Battery and Energy Storage Technology (NY-BEST) Test and Commercialization Center at Eastman Business Park in Rochester, New York, at a cost of $23 million for its almost 1,700 m² laboratory. The center includes the Center for Future Energy Systems, a collaboration between Cornell University of Ithaca, New York and the Rensselaer PolytechniInstitute in Troy, New York. NY-BEST tests, validates and independently certifies diverse forms of energy storage intended for commercial use.

United Kingdom

In the United Kingdom, some fourteen industry and government agencies allied with seven British universities in May 2014 to create the SUPERGEN Energy Storage Hub in order to assist in the coordination of energy storage technology research and development.

Renewable Energy

Wind, solar, and hydroelectricity are three emerging renewable sources of energy

Renewable energy is generally defined as energy that is collected from resources which are naturally replenished on a human timescale, such as sunlight, wind, rain, tides, waves, and geothermal heat. Renewable energy often provides energy in four important areas: electricity generation, air and water heating/cooling, transportation, and rural (off-grid) energy services.

Global public support for energy sources

"Please indicate whether you strongly support, somewhat support, somewhat oppose,
or strongly oppose each way of producing energy"

% very much/somewhat support

Solar	Wind	Hydroelectric	Natural gas	Coal	Nuclear
97	93	91	80	48	38

Source: Ipsos, May 2011

Global publisupport for different energy sources (2011) based on a poll by Ipsos Global @dvisor

Based on REN21's 2016 report, renewables contributed 19.2% to humans' global energy consumption and 23.7% to their generation of electricity in 2014 and 2015, respectively. This energy consumption is divided as 8.9% coming from traditional biomass, 4.2% as heat energy (modern biomass, geothermal and solar heat), 3.9% hydro electricity and 2.2% is electricity from wind, solar, geothermal, and biomass. Worldwide investments in renewable technologies amounted to more than US$286 billion in 2015, with countries like China and the United States heavily investing in wind, hydro, solar and biofuels. Globally, there are an estimated 7.7 million jobs associated with the renewable energy industries, with solar photovoltaics being the largest renewable employer.

Renewable energy resources exist over wide geographical areas, in contrast to other energy sources, which are concentrated in a limited number of countries. Rapid deployment of renewable energy and energy efficiency is resulting in significant energy security, climate change mitigation, and economibenefits. The results of a recent review of the literature concluded that as greenhouse gas (GHG) emitters begin to be held liable for damages resulting from GHG emissions resulting in climate change, a high value for liability mitigation would provide powerful incentives for deployment of renewable energy technologies. In international publiopinion surveys there is strong support for promoting renewable sources such as solar power and wind power. At the national level, at least 30 nations around the world already have renewable energy contributing more than 20 percent of energy supply. National renewable energy markets are projected to continue to grow strongly in the coming decade and beyond. Some places and at least two countries, Iceland and Norway generate all their electricity using renewable energy already, and many other countries have the set a goal to reach 100% renewable energy in the future. For example, in Denmark the government decided to switch the total energy supply (electricity, mobility and heating/cooling) to 100% renewable energy by 2050.

While many renewable energy projects are large-scale, renewable technologies are also suited to rural and remote areas and developing countries, where energy is often crucial in human development. United Nations' Secretary-General Ban Ki-moon has said that renewable energy has the ability to lift the poorest nations to new levels of prosperity. As most of renewables provide electricity, renewable energy deployment is often applied in conjunction with further electrification, which has several benefits: For example, electricity can be converted to heat without losses and even reach higher temperatures than fossil fuels, can be converted into mechanical energy with high efficiency and is clean at the point of consumpion. In addition to that electrification with renewable energy is much more efficient and therefore leads to a significant reduction in primary energy requirements, because most renewables don't have a steam cycle with high losses (fossil power plants usually have losses of 40 to 65%).

Overview

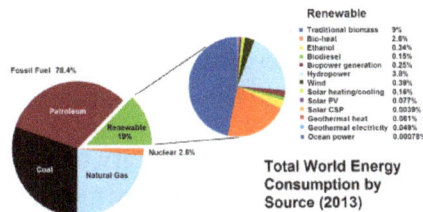

World energy consumption by source. Renewables accounted for 19% in 2012.

PlanetSolar, the world's largest solar-powered boat and the first ever solar electrivehicle to circumnavigate the globe (in 2012)

Renewable energy flows involve natural phenomena such as sunlight, wind, tides, plant growth, and geothermal heat, as the International Energy Agency explains:

Renewable energy is derived from natural processes that are replenished constantly. In its various forms, it derives directly from the sun, or from heat generated deep within the earth. Included in the definition is electricity and heat generated from solar, wind, ocean, hydropower, biomass, geothermal resources, and biofuels and hydrogen derived from renewable resources.

Renewable energy resources and significant opportunities for energy efficiency exist over wide geographical areas, in contrast to other energy sources, which are concentrated in a limited number of countries. Rapid deployment of renewable energy and energy efficiency, and technological diversification of energy sources, would result in significant energy security and economibenefits. It would also reduce environmental pollution such as air pollution caused by burning of fossil fuels and improve publihealth, reduce premature mortalities due to pollution and save associated health costs that amount to several hundred billion dollars annually only in the United States. Renewable energy sources, that derive their energy from the sun, either directly or indirectly, such as hydro and wind, are expected to be capable of supplying humanity energy for almost another 1 billion years, at which point the predicted increase in heat from the sun is expected to make the surface of the earth too hot for liquid water to exist.

Climate change and global warming concerns, coupled with high oil prices, peak oil, and increasing government support, are driving increasing renewable energy legislation, incentives and commercialization. New government spending, regulation and policies helped the industry weather the global financial crisis better than many other sectors. According to a 2011 projection by the International Energy Agency, solar power generators may produce most of the world's electricity within 50 years, reducing the emissions of greenhouse gases that harm the environment.

As of 2011, small solar PV systems provide electricity to a few million households, and micro-hydro configured into mini-grids serves many more. Over 44 million households use biogas made in household-scale digesters for lighting and/or cooking, and more than 166 million households rely on a new generation of more-efficient biomass cookstoves. United Nations' Secretary-General Ban Ki-moon has said that renewable energy has the ability to lift the poorest nations to new levels of prosperity. At the national level, at least 30 nations around the world already have renewable energy contributing more than 20% of energy supply. National renewable energy markets are projected to continue to grow strongly in the coming decade and beyond, and some 120 countries have various policy targets for longer-term shares of renewable energy, including a 20% target of all electricity generated for the European Union by 2020. Some countries have much higher long-term policy targets of up to 100% renewables. Outside Europe, a diverse group of 20 or more other countries target renewable energy shares in the 2020–2030 time frame that range from 10% to 50%.

Renewable energy often displaces conventional fuels in four areas: electricity generation, hot water/space heating, transportation, and rural (off-grid) energy services:

- Power generation

  Renewable hydroelectrienergy provides 16.3% of the worlds electricity. When hydroelectriis combined with other renewables such as wind, geothermal, solar, biomass and waste: together they make the "renewables" total, 21.7% of electricity generation worldwide as of 2013. Renewable power generators are spread across many countries, and wind power alone already provides a significant share of electricity in some areas: for example, 14% in the U.S. state of Iowa, 40% in the northern German state of Schleswig-Holstein, and 49% in Denmark. Some countries get most of their power from renewables, including Iceland (100%), Norway (98%), Brazil (86%), Austria (62%), New Zealand (65%), and Sweden (54%).

- Heating

Solar water heating makes an important contribution to renewable heat in many countries, most notably in China, which now has 70% of the global total (180 GWth). Most of these systems are installed on multi-family apartment buildings and meet a portion of the hot water needs of an estimated 50–60 million households in China. Worldwide, total installed solar water heating systems meet a portion of the water heating needs of over 70 million households. The use of biomass for heating continues to grow as well. In Sweden, national use of biomass energy has surpassed that of oil. Direct geothermal for heating is also growing rapidly. The newest addition to Heating is from Geothermal Heat Pumps which provide both heating and cooling, and also flatten the electridemand curve and are thus an increasing national priority.

- Transportation

A bus fueled by biodiesel

Bioethanol is an alcohol made by fermentation, mostly from carbohydrates produced in sugar or starch crops such as corn, sugarcane, or sweet sorghum. Cellulosibiomass, derived from non-food sources such as trees and grasses is also being developed as a feedstock for ethanol production. Ethanol can be used as a fuel for vehicles in its pure form, but it is usually used as a gasoline additive to increase octane and improve vehicle emissions. Bioethanol is widely used in the USA and in Brazil. Biodiesel can be used as a fuel for vehicles in its pure form, but it is usually used as a diesel additive to reduce levels of particulates, carbon monoxide, and hydrocarbons from diesel-powered vehicles. Biodiesel is produced from oils or fats using transesterification and is the most common biofuel in Europe.

A solar vehicle is an electrivehicle powered completely or significantly by direct solar energy. Usually, photovoltai(PV) cells contained in solar panels convert the sun's energy directly into electrienergy. The term "solar vehicle" usually implies that solar energy is used to power all or part of a vehicle's propulsion. Solar power may be also used to provide power for communications or controls or other auxiliary functions. Solar powered boats have mainly been limited to rivers and canals, but in 2007 an experimental 14m catamaran, the Sun21 sailed the Atlantifrom Seville to Miami, and from there to New York. It was the first crossing of the Atlantipowered only by solar. Solar vehicles are not sold as practical day-to-day transportation devices at present, but are primarily demonstration vehicles and engineering exercises, often sponsored by government agencies. However, indirectly solar-charged vehicles are widespread and solar boats are available commercially.

History

Prior to the development of coal in the mid 19th century, nearly all energy used was renewable. Almost without a doubt the oldest known use of renewable energy, in the form of traditional biomass to fuel fires, dates from 790,000 years ago. Use of biomass for fire did not become commonplace until many hundreds of thousands of years later, sometime between 200,000 and 400,000 years ago. Probably the second oldest usage of renewable energy is harnessing the wind in order to drive ships over water. This practice can be traced back some 7000 years, to ships on the Nile. Moving into the time of recorded history, the primary sources of traditional renewable energy were human labor, animal power, water power, wind, in grain crushing windmills, and firewood, a traditional biomass. A graph of energy use in the United States up until 1900 shows oil and natural gas with about the same importance in 1900 as wind and solar played in 2010.

In the 1860s and '70s there were already fears that civilization would run out of fossil fuels and the need was felt for a better source. In 1873 Professor Augustine Mouchot wrote:

The time will arrive when the industry of Europe will cease to find those natural resources, so necessary for it. Petroleum springs and coal mines are not inexhaustible but are rapidly diminishing in many places. Will man, then, return to the power of water and wind? Or will he emigrate where the most powerful source of heat sends its rays to all? History will show what will come.

In 1885, Werner von Siemens, commenting on the discovery of the photovoltaieffect in the solid state, wrote:

In conclusion, I would say that however great the scientifiimportance of this discovery may be, its practical value will be no less obvious when we reflect that the supply of solar energy is both without limit and without cost, and that it will continue to pour down upon us for countless ages after all the coal deposits of the earth have been exhausted and forgotten.

Max Weber mentioned the end of fossil fuel in the concluding paragraphs of his Die protestantische Ethik und der Geist des Kapitalismus, published in 1905.

Development of solar engines continued until the outbreak of World War I. The importance of solar energy was recognized in a 1911 *ScientifiAmerican* article: "in the far distant future, natural fuels having been exhausted [solar power] will remain as the only means of existence of the human race".

The theory of peak oil was published in 1956. In the 1970s environmentalists promoted the development of renewable energy both as a replacement for the eventual depletion of oil, as well as for an escape from dependence on oil, and the first electricity generating wind turbines appeared. Solar had long been used for heating and cooling, but solar panels were too costly to build solar farms until 1980.

The IEA 2014 World Energy Outlook projects a growth of renewable energy supply from 1,700 gigawatts in 2014 to 4,550 gigawatts in 2040. Fossil fuels received about $550 billion in subsidies in 2013, compared to $120 billion for all renewable energies.

Mainstream Technologies

Wind Power

The 845 MW Shepherds Flat Wind Farm near Arlington, Oregon, USA

Airflows can be used to run wind turbines. Modern utility-scale wind turbines range from around 600 kW to 5 MW of rated power, although turbines with rated output of 1.5–3 MW have become the most common for commercial use; the power available from the wind is a function of the cube of the wind speed, so as wind speed increases, power output increases up to the maximum output for the particular turbine. Areas where winds are stronger and more constant, such as offshore and high altitude sites, are preferred locations for wind farms. Typically full load hours of wind turbines vary between 16 and 57 percent annually, but might be higher in particularly favorable offshore sites.

Globally, the long-term technical potential of wind energy is believed to be five times total current global energy production, or 40 times current electricity demand, assuming all practical barriers needed were overcome. This would require wind turbines to be installed over large areas, particularly in areas of higher wind resources, such as offshore. As offshore wind speeds average ~90% greater than that of land, so offshore resources can contribute substantially more energy than land stationed turbines. In 2014 global wind generation was 706 terawatt-hours or 3% of the worlds total electricity.

Hydropower

The Three Gorges Dam on the Yangtze River in China

In 2015 hydropower generated 16.6% of the worlds total electricity and 70% of all renewable electricity. Since water is about 800 times denser than air, even a slow flowing stream of water, or moderate sea swell, can yield considerable amounts of energy. There are many forms of water energy:

- Historically hydroelectripower came from constructing large hydroelectridams and reservoirs, which are still popular in third world countries. The largest of which is the Three Gorges Dam(2003) in China and the Itaipu Dam(1984) built by Brazil and Paraguay.

- Small hydro systems are hydroelectripower installations that typically produce up to 50 MW of power. They are often used on small rivers or as a low impact development on larger rivers. China is the largest producer of hydroelectricity in the world and has more than 45,000 small hydro installations.

- Run-of-the-river hydroelectricity plants derive kinetienergy from rivers without the creation of a large reservoir. This style of generation may still produce a large amount of electricity, such as the Chief Joseph Dam on the Columbia river in the United States.

Hydropower is produced in 150 countries, with the Asia-Pacifiregion generating 32 percent of global hydropower in 2010. For counties having the largest percentage of electricity from renewables, the top 50 are primarily hydroelectric. China is the largest hydroelectricity producer, with 721 terawatt-hours of production in 2010, representing around 17 percent of domestielectricity use. There are now three hydroelectricity stations larger than 10 GW: the Three Gorges Dam in China, Itaipu Dam across the Brazil/Paraguay border, and Guri Dam in Venezuela.

Wave power, which captures the energy of ocean surface waves, and tidal power, converting the energy of tides, are two forms of hydropower with future potential; however, they are not yet widely employed commercially. A demonstration project operated by the Ocean Renewable Power Company on the coast of Maine, and connected to the grid, harnesses tidal power from the Bay of Fundy, location of world's highest tidal flow. Ocean thermal energy conversion, which uses the temperature difference between cooler deep and warmer surface waters, has currently no economifeasibility.

Solar Energy

Satellite image of the 550-megawatt Topaz Solar Farm in California, USA

Solar energy, radiant light and heat from the sun, is harnessed using a range of ever-evolving technologies such as solar heating, photovoltaics, concentrated solar power (CSP), concentrator photovoltaics (CPV), solar architecture and artificial photosynthesis. Solar technologies are broadly characterized as either passive solar or active solar depending on the way they capture, convert and distribute solar energy. Passive solar techniques include orienting a building to the Sun, selecting materials with favorable thermal mass or light dispersing properties, and designing spaces that naturally circulate air. Active solar technologies encompass solar thermal energy, using solar collectors for heating, and solar power, converting sunlight into electricity either directly using photovoltaics (PV), or indirectly using concentrated solar power (CSP).

A photovoltaisystem converts light into electrical direct current (DC) by taking advantage of the photoelectrieffect. Solar PV has turned into a multi-billion, fast-growing industry, continues to improve its cost-effectiveness, and has the most potential of any renewable technologies together with CSP. Concentrated solar power (CSP) systems use lenses or mirrors and tracking systems to focus a large area of sunlight into a small beam. Commercial concentrated solar power plants were first developed in the 1980s. CSP-Stirling has by far the highest efficiency among all solar energy technologies.

In 2011, the International Energy Agency said that "the development of affordable, inexhaustible and clean solar energy technologies will have huge longer-term benefits. It will increase countries' energy security through reliance on an indigenous, inexhaustible and mostly import-independent resource, enhance sustainability, reduce pollution, lower the costs of mitigating climate change, and keep fossil fuel prices lower than otherwise. These advantages are global. Hence the additional costs of the incentives for early deployment should be considered learning investments; they must be wisely spent and need to be widely shared". In 2014 global solar generation was 186 terawatt-hours, slightly less than 1% of the worlds total grid electricity.

Geothermal Energy

Steam rising from the Nesjavellir Geothermal Power Station in Iceland

High Temperature Geothermal energy is from thermal energy generated and stored in the Earth. Thermal energy is the energy that determines the temperature of matter. Earth's geothermal energy originates from the original formation of the planet and from radioactive decay of minerals (in currently uncertain but possibly roughly equal proportions). The geothermal gradient, which is the difference in temperature between the core of the planet and its surface, drives a continuous conduction of thermal energy in the form of heat from the core to the surface. The adjective *geothermal* originates from the Greek roots *geo*, meaning earth, and *thermos*, meaning heat.

The heat that is used for geothermal energy can be from deep within the Earth, all the way down to Earth's core – 4,000 miles (6,400 km) down. At the core, temperatures may reach over 9,000 °F (5,000 °C). Heat conducts from the core to surrounding rock. Extremely high temperature and pressure cause some rock to melt, which is commonly known as magma. Magma convects upward since it is lighter than the solid rock. This magma then heats rock and water in the crust, sometimes up to 700 °F (371 °C).

From hot springs, geothermal energy has been used for bathing since Paleolithitimes and for space heating since ancient Roman times, but it is now better known for electricity generation.

Low Temperature Geothermal refers to the use of the outer crust of the earth as a Thermal Battery to facilitate Renewable thermal energy for heating and cooling buildings, and other refrigeration and industrial uses. In this form of Geothermal, a Geothermal Heat Pump and Ground-coupled heat exchanger are used together to move heat energy into the earth (for cooling) and out of the earth (for heating) on a varying seasonal basis. Low temperature Geothermal (generally referred to as "GHP") is an increasingly important renewable technology because it both reduces total annual energy loads associated with heating and cooling, and it also flattens the electridemand curve eliminating the extreme summer and winter peak electrisupply requirements. Thus Low Temperature Geothermal/GHP is becoming an increasing national priority with multiple tax credit support and focus as part of the ongoing movement toward Net Zero Energy. New York City has even just passed a law to require GHP anytime is shown to be economical with 20 year financing including the Socialized Cost of Carbon.

Bio Energy

Sugarcane plantation to produce ethanol in Brazil

A CHP power station using wood to supply 30,000 households in France

Biomass is biological material derived from living, or recently living organisms. It most often refers to plants or plant-derived materials which are specifically called lignocellulosibiomass. As an energy source, biomass can either be used directly via combustion to produce heat, or indirectly after converting it to various forms of biofuel. Conversion of biomass to biofuel can be achieved by different methods which are broadly classified into: *thermal*, *chemical*, and *biochemical* methods. Wood remains the largest biomass energy source today; examples include forest residues – such as dead trees, branches and tree stumps –, yard clippings, wood chips and even municipal solid waste. In the second sense, biomass includes plant or animal matter that can be converted into fibers or other industrial chemicals, including biofuels. Industrial biomass can be grown from numerous types of plants, including miscanthus, switchgrass, hemp, corn, poplar, willow, sorghum, sugarcane, bamboo, and a variety of tree species, ranging from eucalyptus to oil palm (palm oil).

Plant energy is produced by crops specifically grown for use as fuel that offer high biomass output per hectare with low input energy. Some examples of these plants are wheat, which typically yield 7.5–8 tonnes of grain per hectare, and straw, which typically yield 3.5–5 tonnes per hectare in the UK. The grain can be used for liquid transportation fuels while the straw can be burned to produce heat or electricity. Plant biomass can also be degraded from cellulose to glucose through a series of chemical treatments, and the resulting sugar can then be used as a first generation biofuel.

Biomass can be converted to other usable forms of energy like methane gas or transportation fuels like ethanol and biodiesel. Rotting garbage, and agricultural and human waste, all release methane gas – also called landfill gas or biogas. Crops, such as corn and sugarcane, can be fermented to produce the transportation fuel, ethanol. Biodiesel, another transportation fuel, can be produced from left-over food products like vegetable oils and animal fats. Also, biomass to liquids (BTLs) and cellulosiethanol are still under research. There is a great deal of research involving algal fuel or algae-derived biomass due to the fact that it's a non-food resource and can be produced at rates 5 to 10 times those of other types of land-based agriculture, such as corn and soy. Once harvested, it can be fermented to produce biofuels such as ethanol, butanol, and methane, as well as biodiesel and hydrogen. The biomass used for electricity generation varies by region. Forest by-products, such as wood residues, are common in the United States. Agricultural waste is common in Mauritius (sugar cane residue) and Southeast Asia (rice husks). Animal husbandry residues, such as poultry litter, are common in the United Kingdom.

Biofuels include a wide range of fuels which are derived from biomass. The term covers solid, liquid, and gaseous fuels. Liquid biofuels include bioalcohols, such as bioethanol, and oils, such as biodiesel. Gaseous biofuels include biogas, landfill gas and synthetigas. Bioethanol is an alcohol made by fermenting the sugar components of plant materials and it is made mostly from sugar and starch crops. These include maize, sugarcane and, more recently, sweet sorghum. The latter crop is particularly suitable for growing in dryland conditions, and is being investigated by International Crops Research Institute for the Semi-Arid Tropics for its potential to provide fuel, along with food and animal feed, in arid parts of Asia and Africa.

With advanced technology being developed, cellulosibiomass, such as trees and grasses, are also used as feedstocks for ethanol production. Ethanol can be used as a fuel for vehicles in its pure form, but it is usually used as a gasoline additive to increase octane and improve vehicle emissions. Bioethanol is widely used in the United States and in Brazil. The energy costs for producing bio-ethanol are almost equal to, the energy yields from bio-ethanol. However, according to the European Environment Agency, biofuels do not address global warming concerns. Biodiesel is made from vegetable oils, animal fats or recycled greases. It can be used as a fuel for vehicles in its pure form, or more commonly as a diesel additive to reduce levels of particulates, carbon monoxide, and hydrocarbons from diesel-powered vehicles. Biodiesel is produced from oils or fats using transesterification and is the most common biofuel in Europe. Biofuels provided 2.7% of the world's transport fuel in 2010.

Biomass, biogas and biofuels are burned to produce heat/power and in doing so harm the environment. Pollutants such as sulphurous oxides (SO_x), nitrous oxides (NO_x), and particulate matter (PM) are produced from the combustion of biomass; the World Health Organisation estimates that 7 million premature deaths are caused each year by air pollution. Biomass combustion is a major contributor. The life cycle of the plants is sustainable, the lives of people less so.

Energy Storage

Energy storage is a collection of methods used to store electrical energy on an electrical power grid, or off it. Electrical energy is stored during times when production (especially from intermittent power plants such as renewable electricity sources such as wind power, tidal power, solar power) exceeds consumption, and returned to the grid when production falls below consumption. Water pumped into a hydroelectridam is the largest form of power storage.

Market and Industry Trends

Growth of Renewables

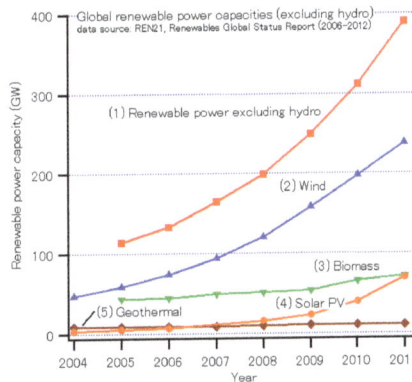

Global growth of renewables through 2011

From the end of 2004, worldwide renewable energy capacity grew at rates of 10–60% annually for many technologies. For wind power and many other renewable technologies, growth accelerated in 2009 relative to the previous four years. More wind power capacity was added during 2009 than any other renewable technology. However, grid-connected PV increased the fastest of all renewables technologies, with a 60% annual average growth rate. In 2010, renewable power constituted about a third of the newly built power generation capacities.

Projections vary, but scientists have advanced a plan to power 100% of the world's energy with wind, hydroelectric, and solar power by the year 2030.

According to a 2011 projection by the International Energy Agency, solar power generators may produce most of the world's electricity within 50 years, reducing the emissions of greenhouse gases that harm the environment. CedriPhilibert, senior analyst in the renewable energy division at the IEA said: "Photovoltaiand solar-thermal plants may meet most of the world's demand for electricity by 2060 – and half of all energy needs – with wind, hydropower and biomass plants supplying much of the remaining generation". "Photovoltaiand concentrated solar power together can become the major source of electricity", Philibert said.

In 2014 global wind power capacity expanded 16% to 369,553 MW. Yearly wind energy production is also growing rapidly and has reached around 4% of worldwide electricity usage, 11.4% in the EU, and it is widely used in Asia, and the United States. In 2015, worldwide installed photovoltaics capacity increased to 227 gigawatts (GW), sufficient to supply 1 percent of global electricity demands. Solar thermal energy stations operate in the USA and Spain, and as of 2016, the largest of these is the 392 MW Ivanpah Solar ElectriGenerating System in California. The world's largest geothermal power installation is The Geysers in California, with a rated capacity of 750 MW. Brazil has one of the largest renewable energy programs in the world, involving production of ethanol fuel from sugar cane, and ethanol now provides 18% of the country's automotive fuel. Ethanol fuel is also widely available in the USA.

Selected renewable energy global indicators	2008	2009	2010	2011	2012	2013	2014	2015
Investment in new renewable capacity (annual) (10⁹ USD)	182	178	237	279	256	232	270	285
Renewables power capacity (existing) (GWe)	1,140	1,230	1,320	1,360	1,470	1,578	1,712	1,849
Hydropower capacity (existing) (GWe)	885	915	945	970	990	1,018	1,055	1,064
Wind power capacity (existing) (GWe)	121	159	198	238	283	319	370	433
Solar PV capacity (grid-connected) (GWe)	16	23	40	70	100	138	177	227
Solar hot water capacity (existing) (GWth)	130	160	185	232	255	373	406	435
Ethanol production (annual) (10⁹ litres)	67	76	86	86	83	87	94	98
Biodiesel production (annual) (10⁹ litres)	12	17.8	18.5	21.4	22.5	26	29.7	30
Countries with policy targets for renewable energy use	79	89	98	118	138	144	164	173
Source: The Renewable Energy Policy Network for the 21st Century (REN21)–Global Status Report								

EconomiTrends

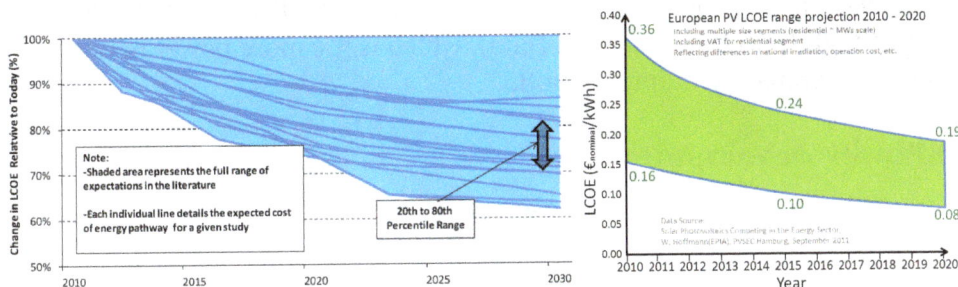

Projection of levelized cost for wind in the U.S. (left) and solar power in Europe

Renewable energy technologies are getting cheaper, through technological change and through the benefits of mass production and market competition. A 2011 IEA report said: "A portfolio of renewable energy technologies is becoming cost-competitive in an increasingly broad range of circumstances, in some cases providing investment opportunities without the need for specificeconomisupport," and added that "cost reductions in critical technologies, such as wind and solar, are set to continue."

Hydro-electricity and geothermal electricity produced at favourable sites are now the cheapest way to generate electricity. Renewable energy costs continue to drop, and the levelised cost of electricity (LCOE) is declining for wind power, solar photovoltai(PV), concentrated solar power (CSP) and some biomass technologies. Renewable energy is also the most economisolution for new grid-connected capacity in areas with good resources. As the cost of renewable power falls, the scope of economically viable applications increases. Renewable technologies are now often the most economisolution for new generating capacity. Where "oil-fired generation is the predominant power generation source (e.g. on islands, off-grid and in some countries) a lower-cost renewable solution almost always exists today". A series of studies by the US National Renewable Energy Laboratory modeled the "grid in the Western US under a number of different scenarios where intermittent renewables accounted for 33 percent of the total power." In the models, inefficiencies in cycling the fossil fuel plants to compensate for the variation in solar and wind energy resulted in an additional cost of "between $0.47 and $1.28 to each MegaWatt hour generated"; however, the savings in the cost of the fuels saved "adds up to $7 billion, meaning the added costs are, at most, two percent of the savings."

Hydroelectricity

Only 25% of the worlds estimated hydroelectripotential of 14,000 TWh/year has been developed, with Africa, Asia and Latin America having the greatest potential. The Three Gorges Dam in Hubei, China, has the world's largest instantaneous generating capacity (22,500 MW), with the Itaipu Dam in Brazil/Paraguay in second place (14,000 MW). The Three Gorges Dam is operated jointly with the much smaller Gezhouba Dam (3,115

MW). As of 2012, the total generating capacity of this two-dam complex is 25,615 MW. In 2008, this complex generated 98 TWh of electricity (81 TWh from the Three Gorges Dam and 17 TWh from the Gezhouba Dam), which is 3% more power in one year than the 95 TWh generated by Itaipu in 2008.

Wind Power Development

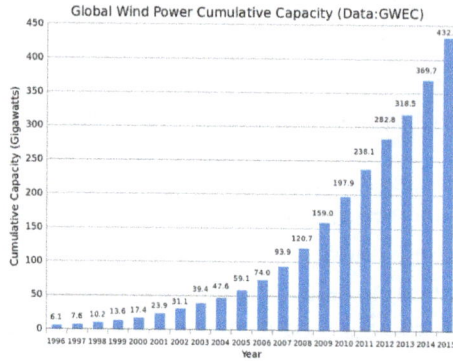

Worldwide growth of wind capacity (1996–2014)

Four offshore wind farms are in the Thames Estuary area: Kentish Flats, Gunfleet Sands, Thanet and London Array. The latter is the largest in the world as of April 2013.

Wind power is widely used in Europe, China, and the United States. From 2004 to 2014, worldwide installed capacity of wind power has been growing from 47 GW to 369 GW—a more than sevenfold increase within 10 years with 2014 breaking a new record in global installations (51 GW). As of the end of 2014, China, the United States and Germany combined accounted for half of total global capacity. Several other countries have achieved relatively high levels of wind power penetration, such as 21% of stationary electricity production in Denmark, 18% in Portugal, 16% in Spain, and 14% in Ireland in 2010 and have since continued to expand their installed capacity. More than 80 countries around the world are using wind power on a commercial basis.

- Offshore wind power

    As of 2014, offshore wind power amounted to 8,771 megawatt of global installed capacity. Although offshore capacity doubled within three years (from 4,117 MW in 2011), it accounted for only 2.3% of the total wind power capacity. The United Kingdom is the undisputed leader of offshore power with half of the world's installed capacity ahead of Denmark, Germany, Belgium and China.

- List of offshore and onshore wind farms

As of 2012, the Alta Wind Energy Center (California, 1,020 MW) is the world's largest wind farm. The London Array (630 MW) is the largest offshore wind farm in the world. The United Kingdom is the world's leading generator of offshore wind power, followed by Denmark. There are several large offshore wind farms operational and under construction and these include Anholt (400 MW), BARD (400 MW), Clyde (548 MW), Fântânele-Cogeala(600 MW), Greater Gabbard (500 MW), Lincs (270 MW), London Array (630 MW), Lower Snake River (343 MW), Macarthur (420 MW), Shepherds Flat (845 MW), and the Sheringham Shoal (317 MW).

Solar Thermal

The 377 MW Ivanpah Solar ElectriGenerating System with all three towers under load, Feb 2014. Taken from I-15.

Solar Towers of the PS10 and PS20 solar thermal plants in Spain

The United States conducted much early research in photovoltaics and concentrated solar power. The U.S. is among the top countries in the world in electricity generated by the Sun and several of the world's largest utility-scale installations are located in the desert Southwest.

The oldest solar thermal power plant in the world is the 354 megawatt (MW) SEGS thermal power plant, in California. The Ivanpah Solar ElectriGenerating System is a solar thermal power project in the California Mojave Desert, 40 miles (64 km) southwest of Las Vegas, with a gross capacity of 377 MW. The 280 MW Solana Generating Station is a solar power plant near Gila Bend, Arizona, about 70 miles (110 km) southwest of Phoenix, completed in 2013. When commissioned it was the largest parabolitrough plant in the world and the first U.S. solar plant with molten salt thermal energy storage.

The solar thermal power industry is growing rapidly with 1.3 GW under construction in 2012 and more planned. Spain is the epicenter of solar thermal power development with 873 MW under construction, and a further 271 MW under development. In the United States, 5,600 MW of solar thermal power projects have been announced. Several power plants have been constructed in the Mojave Desert, Southwestern United States. The Ivanpah Solar Power Facility being the most recent. In developing countries, three World Bank projects for integrated solar thermal/combined-cycle gas-turbine power plants in Egypt, Mexico, and Morocco have been approved.

Photovoltaics (PV) uses solar cells assembled into solar panels to convert sunlight into electricity. It's a fast-growing technology doubling its worldwide installed capacity every couple of years. PV systems range from small, residential and commercial rooftop or building integrated installations, to large utility-scale photovoltaipower station. The predominant PV technology is crystalline silicon, while thin-film solar cell technology accounts for about 10 percent of global photovoltaideployment. In recent years, PV technology has improved its electricity generating efficiency, reduced the installation cost per watt as well as its energy payback time, and has reached grid parity in at least 30 different markets by 2014. Financial institutions are predicting a second solar "gold rush" in the near future.

At the end of 2014, worldwide PV capacity reached at least 177,000 megawatts. Photovoltaics grew fastest in China, followed by Japan and the United States, while Germany remains the world's largest overall producer of photovoltaipower, contributing about 7.0 percent to the overall electricity generation. Italy meets 7.9 percent of its electricity demands with photovoltaipower—the highest share worldwide. For 2015, global cumulative capacity is forecasted to increase by more than 50 gigawatts (GW). By 2018, worldwide capacity is projected to reach as much as 430 gigawatts. This corresponds to a tripling within five years. Solar power is forecasted to become the world's largest source of electricity by 2050, with solar photovoltaics and concentrated solar power contributing 16% and 11%, respectively. This requires an increase of installed PV capacity to 4,600 GW, of which more than half is expected to be deployed in China and India.

Photovoltaic Power Stations

Commercial concentrated solar power plants were first developed in the 1980s. As the cost of solar electricity has fallen, the number of grid-connected solar PV systems has grown into the millions and utility-scale solar power stations with hundreds of megawatts are being built. Solar PV is rapidly becoming an inexpensive, low-carbon technology to harness renewable energy from the Sun.

Solar panels at the 550 MW Topaz Solar Farm

Nellis Solar Power Plant, photovoltaipower plant in Nevada, USA

Many solar photovoltaipower stations have been built, mainly in Europe, China and the USA. The 579 MW Solar Star, in the United States, is the world's largest PV power station.

Many of these plants are integrated with agriculture and some use tracking systems that follow the sun's daily path across the sky to generate more electricity than fixed-mounted systems. There are no fuel costs or emissions during operation of the power stations.

However, when it comes to renewable energy systems and PV, it is not just large systems that matter. Building-integrated photovoltaics or "onsite" PV systems use existing land and structures and generate power close to where it is consumed.

Biofuel Development

Brazil produces bioethanol made from sugarcane available throughout the country. A typical gas station with dual fuel service is marked "A" for alcohol (ethanol) and "G" for gasoline.

Biofuels provided 3% of the world's transport fuel in 2010. Mandates for blending bio-fuels exist in 31 countries at the national level and in 29 states/provinces. According to the International Energy Agency, biofuels have the potential to meet more than a quarter of world demand for transportation fuels by 2050.

Since the 1970s, Brazil has had an ethanol fuel program which has allowed the country to become the world's second largest producer of ethanol (after the United States) and the world's largest exporter. Brazil's ethanol fuel program uses modern equipment and cheap sugarcane as feedstock, and the residual cane-waste (bagasse) is used to produce heat and power. There are no longer light vehicles in Brazil running on pure gasoline. By the end of 2008 there were 35,000 filling stations throughout Brazil with at least one ethanol pump. Unfortunately, Operation Car Wash has seriously eroded publitrust in oil companies and has implicated several high ranking Brazilian officials.

Nearly all the gasoline sold in the United States today is mixed with 10% ethanol, and motor vehicle manufacturers already produce vehicles designed to run on much higher ethanol blends. Ford, Daimler AG, and GM are among the automobile companies that sell "flexible-fuel" cars, trucks, and minivans that can use gasoline and ethanol blends ranging from pure gasoline up to 85% ethanol. By mid-2006, there were approximately 6 million ethanol compatible vehicles on U.S. roads.

Geothermal Development

Geothermal plant at The Geysers, California, USA

Geothermal power is cost effective, reliable, sustainable, and environmentally friendly, but has historically been limited to areas near tectoniplate boundaries. Recent technological advances have expanded the range and size of viable resources, especially for applications such as home heating, opening a potential for widespread exploitation. Geothermal wells release greenhouse gases trapped deep within the earth, but these emissions are much lower per energy unit than those of fossil fuels. As a result, geothermal power has the potential to help mitigate global warming if widely deployed in place of fossil fuels.

The International Geothermal Association (IGA) has reported that 10,715 MW of geo-thermal power in 24 countries is online, which is expected to generate 67,246 GWh of electricity in 2010. This represents a 20% increase in geothermal power online capacity since 2005. IGA projects this will grow to 18,500 MW by 2015, due to the large number of projects presently under consideration, often in areas previously assumed to have little exploitable resource.

In 2010, the United States led the world in geothermal electricity production with 3,086 MW of installed capacity from 77 power plants; the largest group of geothermal power plants in the world is located at The Geysers, a geothermal field in California. The Philippines follows the US as the second highest producer of geothermal power in the world, with 1,904 MW of capacity online; geothermal power makes up approximately 18% of the country's electricity generation.

Developing Countries

Renewable energy technology has sometimes been seen as a costly luxury item by crit-ics, and affordable only in the affluent developed world. This erroneous view has per-sisted for many years, but 2015 was the first year when investment in non-hydro renew-ables, was higher in developing countries, with $156 billion invested, mainly in China, India, and Brazil.

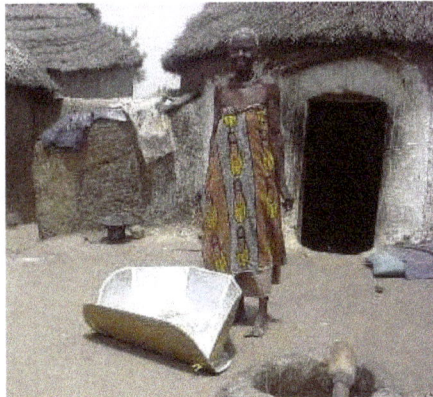

Solar cookers use sunlight as energy source for outdoor cooking

Renewable energy can be particularly suitable for developing countries. In rural and re-mote areas, transmission and distribution of energy generated from fossil fuels can be difficult and expensive. Producing renewable energy locally can offer a viable alternative.

Technology advances are opening up a huge new market for solar power: the approximate-ly 1.3 billion people around the world who don't have access to grid electricity. Even though they are typically very poor, these people have to pay far more for lighting than people in rich countries because they use inefficient kerosene lamps. Solar power costs half as much as lighting with kerosene. As of 2010, an estimated 3 million households get power from small solar PV systems. Kenya is the world leader in the number of solar power systems installed per capita. More than 30,000 very small solar panels, each producing 12 to 30 watts, are sold in Kenya annually. Some Small Island Developing States (SIDS) are also turning to solar power to reduce their costs and increase their sustainability.

Micro-hydro configured into mini-grids also provide power. Over 44 million households use biogas made in household-scale digesters for lighting and/or cooking, and more than 166 million households rely on a new generation of more-efficient biomass cookstoves. Clean liquid fuel sourced from renewable feedstocks are used for cooking and lighting in energy-poor areas of the developing world. Alcohol fuels (ethanol and methanol) can be produced sustainably from non-food sugary, starchy, and cellulostifeedstocks. Project Gaia, Inc. and CleanStar Mozambique are implementing clean cooking programs with liquid ethanol stoves in Ethiopia, Kenya, Nigeria and Mozambique.

Renewable energy projects in many developing countries have demonstrated that renewable energy can directly contribute to poverty reduction by providing the energy needed for creating businesses and employment. Renewable energy technologies can also make indirect contributions to alleviating poverty by providing energy for cooking, space heating, and lighting. Renewable energy can also contribute to education, by providing electricity to schools.

Industry and Policy Trends

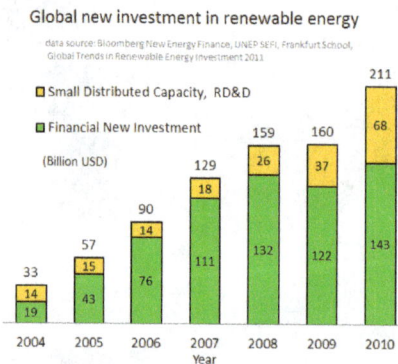

Global new investment in renewable energy

data source: Bloomberg New Energy Finance, UNEP SEFI, Frankfurt School, Global Trends in Renewable Energy Investment 2011

☐ Small Distributed Capacity, RD&D

☐ Financial New Investment

(Billion USD)

Year	2004	2005	2006	2007	2008	2009	2010
Total	33	57	90	129	159	160	211
Small Distributed Capacity, RD&D	14	15	14	18	26	37	68
Financial New Investment	19	43	76	111	132	122	143

Global New Investments in Renewable Energy

U.S. President Barack Obama's American Recovery and Reinvestment Act of 2009 includes more than $70 billion in direct spending and tax credits for clean energy and associated transportation programs. Leading renewable energy companies include First Solar, Gamesa, GE Energy, Hanwha Q Cells, Sharp Solar, Siemens, SunOpta, Suntech Power, and Vestas.

Many national, state, and local governments have also created green banks. A green bank is a quasi-publifinancial institution that uses publicapital to leverage private investment in clean energy technologies. Green banks use a variety of financial tools to bridge market gaps that hinder the deployment of clean energy.

The military has also focused on the use of renewable fuels for military vehicles. Unlike fossil fuels, renewable fuels can be produced in any country, creating a strategiadvantage. The US military has already committed itself to have 50% of its energy consumption come from alternative sources.

The International Renewable Energy Agency (IRENA) is an intergovernmental organization for promoting the adoption of renewable energy worldwide. It aims to provide concrete policy advice and facilitate capacity building and technology transfer. IRENA was formed on 26 January 2009, by 75 countries signing the charter of IRENA. As of March 2010, IRENA has 143 member states who all are considered as founding members, of which 14 have also ratified the statute.

As of 2011, 119 countries have some form of national renewable energy policy target or renewable support policy. National targets now exist in at least 98 countries. There is also a wide range of policies at state/provincial and local levels.

United Nations' Secretary-General Ban Ki-moon has said that renewable energy has the ability to lift the poorest nations to new levels of prosperity. In October 2011, he "announced the creation of a high-level group to drum up support for energy access, energy efficiency and greater use of renewable energy. The group is to be co-chaired by Kandeh Yumkella, the chair of UN Energy and director general of the UN Industrial Development Organisation, and Charles Holliday, chairman of Bank of America".

100% Renewable Energy

The incentive to use 100% renewable energy, for electricity, transport, or even total primary energy supply globally, has been motivated by global warming and other ecological as well as economicconcerns. The Intergovernmental Panel on Climate Change has said that there are few fundamental technological limits to integrating a portfolio of renewable energy technologies to meet most of total global energy demand. Renewable energy use has grown much faster than even advocates anticipated. At the national level, at least 30 nations around the world already have renewable energy contributing more than 20% of energy supply. Also, Professors S. Pacala and Robert H. Socolow have developed a series of "stabilization wedges" that can allow us to maintain our quality of life while avoiding catastrophiclimate change, and "renewable energy sources," in aggregate, constitute the largest number of their "wedges."

Using 100% renewable energy was first suggested in a Science paper published in 1975 by Danish physicist Bent Sørensen. It was followed by several other proposals, until in 1998 the first detailed analysis of scenarios with very high shares of renewables were published. These were followed by the first detailed 100% scenarios. In 2006 a PhD thesis was published by Czisch in which it was shown that in a 100% renewable scenario energy supply could match demand in every hour of the year in Europa and North Africa. In the same year Danish Energy professor Henrik Lund published a first paper in which he addresses the optimal combination of renewables, which was followed by several other papers on the transition to 100% renewable energy in Denmark. Since then Lund has been publishing several papers on 100% renewable energy. After 2009 publications began to rise steeply, covering 100% scenarios for countries in Europa, America, Australia and other parts of the world.

In 2011 Mark Z. Jacobson, professor of civil and environmental engineering at Stanford University, and Mark Delucchi published a study on 100% renewable global energy supply in the journal Energy Policy. They found producing all new energy with wind power, solar power, and hydropower by 2030 is feasible and existing energy supply arrangements could be replaced by 2050. Barriers to implementing the renewable energy plan are seen to be "primarily social and political, not technological or economic". They also found that energy costs with a wind, solar, water system should be similar to today's energy costs.

Similarly, in the United States, the independent National Research Council has noted that "sufficient domestirenewable resources exist to allow renewable electricity to play a significant role in future electricity generation and thus help confront issues related to climate change, energy security, and the escalation of energy costs ... Renewable energy is an attractive option because renewable resources available in the United States, taken collectively, can supply significantly greater amounts of electricity than the total current or projected domestidemand." .

The most significant barriers to the widespread implementation of large-scale renewable energy and low carbon energy strategies are primarily political and not technological. According to the 2013 *Post Carbon Pathways* report, which reviewed many international studies, the key roadblocks are: climate change denial, the fossil fuels lobby, political inaction, unsustainable energy consumption, outdated energy infrastructure, and financial constraints.

Emerging Technologies

Other renewable energy technologies are still under development, and include cellulosiethanol, hot-dry-rock geothermal power, and marine energy. These technologies are not yet widely demonstrated or have limited commercialization. Many are on the horizon and may have potential comparable to other renewable energy technologies, but still depend on attracting sufficient attention and research, development and demonstration (RD&D) funding.

There are numerous organizations within the academic, federal, and commercial sectors conducting large scale advanced research in the field of renewable energy. This research spans several areas of focus across the renewable energy spectrum. Most of the research is targeted at improving efficiency and increasing overall energy yields. Multiple federally supported research organizations have focused on renewable energy in recent years. Two of the most prominent of these labs are Sandia National Laboratories and the National Renewable Energy Laboratory (NREL), both of which are funded by the United States Department of Energy and supported by various corporate partners. Sandia has a total budget of $2.4 billion while NREL has a budget of $375 million.

- Enhanced geothermal system

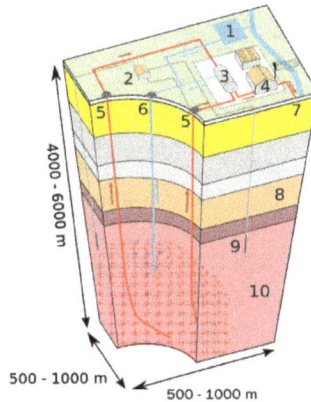

Enhanced geothermal system

Enhanced geothermal systems (EGS) are a new type of geothermal power technologies that do not require natural convective hydrothermal resources. The vast majority of geothermal energy within drilling reach is in dry and non-porous rock. EGS technologies "enhance" and/or create geothermal resources in this "hot dry rock (HDR)" through hydraulistimulation. EGS and HDR technologies, like hydrothermal geothermal, are expected to be baseload resources which produce power 24 hours a day like a fossil plant. Distinct from hydrothermal, HDR and EGS may be feasible anywhere in the world, depending on the economilimits of drill depth. Good locations are over deep granite covered by a thick (3–5 km) layer of insulating sediments which slow heat loss. There are HDR and EGS systems currently being developed and tested in France, Australia, Japan, Germany, the U.S. and Switzerland. The largest EGS project in the world is a 25 megawatt demonstration plant currently being developed in the Cooper Basin, Australia. The Cooper Basin has the potential to generate 5,000–10,000 MW.

- Cellulosiethanol

Several refineries that can process biomass and turn it into ethano are built by companies such as Iogen, POET, and Abengoa, while other companies such as the Verenium Corporation, Novozymes, and DyadiInternational are producing enzymes which could enable future commercialization. The shift from food crop feedstocks to waste residues and native grasses offers significant opportunities for a range of players, from farmers to biotechnology firms, and from project developers to investors.

- Marine energy

Rance Tidal Power Station, France

Marine energy (also sometimes referred to as ocean energy) refers to the energy carried by ocean waves, tides, salinity, and ocean temperature differences. The movement of water in the world's oceans creates a vast store of kinetienergy, or energy in motion. This energy can be harnessed to generate electricity to power homes, transport and industries. The term marine energy encompasses both wave power – power from surface waves, and tidal power – obtained from the kinetienergy of large bodies of moving water. Reverse electrodialysis (RED) is a technology for generating electricity by mixing fresh river water and salty sea water in large power cells designed for this purpose; as of 2016 it is being tested at a small scale (50 kW). Offshore wind power is not a form of marine energy, as wind power is derived from the wind, even if the wind turbines are placed over water. The oceans have a tremendous amount of energy and are close to many if not most concentrated populations. Ocean energy has the potential of providing a substantial amount of new renewable energy around the world.

#	Station	Country	Location	Capacity	Refs
1.	Sihwa Lake Tidal Power Station	South Korea	37°18′47″N 126°36′46″E37.31306°N 126.61278°E	254 MW	
2.	Rance Tidal Power Station	France	48°37′05″N 02°01′24″W48.61806°N 2.02333°W	240 MW	
3.	Annapolis Royal Generating Station	Canada	44°45′07″N 65°30′40″W44.75194°N 65.51111°W	20 MW	

Experimental solar power

Concentrated photovoltaics (CPV) systems employ sunlight concentrated onto photovoltaisurfaces for the purpose of electricity generation. Thermoelectric, or "thermovoltaic" devices convert a temperature difference between dissimilar materials into an electricurrent.

- Floating solar arrays

  Floating solar arrays are PV systems that float on the surface of drinking water reservoirs, quarry lakes, irrigation canals or remediation and tailing ponds. A small number of such systems exist in France, India, Japan, South Korea, the United Kingdom, Singapore and the United States. The systems are said to have advantages over photovoltaics on land. The cost of land is more expensive, and there are fewer rules and regulations for structures built on bodies of water not used for recreation. Unlike most land-based solar plants, floating arrays can be unobtrusive because they are hidden from publiview. They achieve higher efficiencies than PV panels on land, because water cools the panels. The panels have a special coating to prevent rust or corrosion. In May 2008, the Far Niente Winery in Oakville, California, pioneered the world's first floatovoltaisystem by installing 994 solar PV modules with a total capacity of 477 kW onto 130 pontoons and floating them on the winery's irrigation pond. Utility-scale floating PV farms are starting to be built. Kyocera will develop the world's largest, a 13.4 MW farm on the reservoir above Yamakura Dam in Chiba Prefecture using 50,000 solar panels. Salt-water resistant floating farms are also being constructed for ocean use. The largest so far announced floatovoltaiproject is a 350 MW power station in the Amazon region of Brazil.

- Solar-assisted heat pump

  A heat pump is a device that provides heat energy from a source of heat to a destination called a "heat sink". Heat pumps are designed to move thermal energy opposite to the direction of spontaneous heat flow by absorbing heat from a cold space and releasing it to a warmer one. A solar-assisted heat pump represents the integration of a heat pump and thermal solar panels in a single integrated system. Typically these two technologies are used separately (or only placing them in parallel) to produce hot water. In this system the solar thermal panel performs the function of the low temperature heat source and the heat produced is used to feed the heat pump's evaporator. The goal of this system is to get high COP and then produce energy in a more efficient and less expensive way.

  It is possible to use any type of solar thermal panel (sheet and tubes, roll-bond, heat pipe, thermal plates) or hybrid (mono/polycrystalline, thin film) in combination with the heat pump. The use of a hybrid panel is preferable because it allows to cover a part of the electricity demand of the heat pump and reduce the power consumption and consequently the variable costs of the system.

- Artificial photosynthesis

  Artificial photosynthesis uses techniques including nanotechnology to store solar electromagnetienergy in chemical bonds by splitting water to produce hydrogen and then using carbon dioxide to make methanol. Researchers in this field are striving to design molecular mimics of photosynthesis that utilize a wider region of the solar spectrum, employ catalytisystems made from abundant, inexpensive materials that are robust, readily repaired, non-toxic, stable in a variety of environmental conditions and perform more efficiently allowing a greater proportion of photon energy to end up in the storage compounds, i.e., carbohydrates (rather than building and sustaining living cells). However, prominent research faces hurdles, Sun Catalytix a MIT spin-off stopped scaling up their prototype fuel-cell in 2012, because it offers few savings over other ways to make hydrogen from sunlight.

- Algae fuels

Producing liquid fuels from oil-rich varieties of algae is an ongoing research topic. Various microalgae grown in open or closed systems are being tried including some system that can be set up in brownfield and desert lands.

- Space-based solar power

For either photovoltaior thermal systems, one option is to loft them into space, particularly Geosynchronous orbit. To be competitive with Earth-based solar power systems, the specifimass (kg/kW) times the cost to loft mass plus the cost of the parts needs to be $2400 or less. I.e., for a parts cost plus rectenna of $1100/kW, the product of the $/kg and kg/kW must be $1300/kW or less. Thus for 6.5 kg/kW, the transport cost cannot exceed $200/kg. While that will require a 100 to one reduction, SpaceX is targeting a ten to one reduction, Reaction Engines may make a 100 to one reduction possible.

- Carbon-neutral and negative fuels

Carbon-neutral fuels are synthetifuels (including methane, gasoline, diesel fuel, jet fuel or ammonia) produced by hydrogenating waste carbon dioxide recycled from power plant flue-gas emissions, recovered from automotive exhaust gas, or derived from carboniacid in seawater. Such fuels are considered carbon-neutral because they do not result in a net increase in atmospherigreenhouse gases.

Carbon-neutral fuels offer relatively low cost energy storage, alleviating the problems of wind and solar variability, and they enable distribution of wind, water, and solar power through existing natural gas pipelines. Nighttime wind power is considered the most economical form of electrical power with which to synthesize fuel, because the load curve for electricity peaks sharply during the warmest hours of the day, but wind tends to blow slightly more at night than during the day, so, the price of nighttime wind power is often much less expensive than any alternative. Germany has built a 250 kilowatt synthetimethane plant which they are scaling up to 10 megawatts.

The George Olah carbon dioxide recycling plant in Grindavík, Iceland has been producing 2 million liters of methanol transportation fuel per year from flue exhaust of the Svartsengi Power Station since 2011. It has the capacity to produce 5 million liters per year.

Debate

Renewable electricity production, from sources such as wind power and solar power, is sometimes criticized for being variable or intermittent, but is not true for concentrated solar, geothermal and biofuels, that have continuity. In any case, the International Energy Agency has stated that deployment of renewable technologies usually increases the diversity of electricity sources and, through local generation, contributes to the flexibility of the system and its resistance to central shocks.

There have been "not in my back yard" (NIMBY) concerns relating to the visual and other impacts of some wind farms, with local residents sometimes fighting or blocking construction. In the USA, the Massachusetts Cape Wind project was delayed for years partly because of aestheticconcerns. However, residents in other areas have been more positive. According to a town councilor, the overwhelming majority of locals believe that the Ardrossan Wind Farm in Scotland has enhanced the area.

A recent UK Government document states that "projects are generally more likely to succeed if they have broad publisupport and the consent of local communities. This means giving communities both a say and a stake". In countries such as Germany and Denmark many renewable projects are owned by communities, particularly through cooperative structures, and contribute significantly to overall levels of renewable energy deployment.

The market for renewable energy technologies has continued to grow. Climate change concerns and increasing in green jobs, coupled with high oil prices, peak oil, oil wars, oil spills, promotion of electrivehicles and renewable electricity, nuclear disasters and increasing government support, are driving increasing renewable energy legislation, incentives and commercialization. New government spending, regulation and policies helped the industry weather the 2009 economicrisis better than many other sectors.

References

- Hanisch, Christian; Diekmann, Jan; Stieger, Alexander; Haselrieder, Wolfgang; Kwade, Arno (2015). "27". In Yan, Jinyue; Cabeza, Luisa F.; Sioshansi, Ramteen. Handbook of Clean Energy Systems - Recycling of Lithium-Ion Batteries (5 Energy Storage ed.). John Wiley & Sons, Ltd. pp. 2865–2888. ISBN 9781118991978.

- David Linden, Thomas B. Reddy (ed). Handbook Of Batteries 3rd Edition. McGraw-Hill, New York, 2002 ISBN 0-07-135978-8.

- Katerina E. Aifantis et al, High Energy Density Lithium Batteries: Materials, Engineering, Applications Wiley-VCH, 2010 ISBN 3-527-32407-0 page 66

- Aifantis, Katerina E.; Hackney, Stephen A.; Kumar, R. Vasant (March 30, 2010). High Energy Density Lithium Batteries: Materials, Engineering, Applications. John Wiley & Sons. ISBN 978-3-527-63002-8.

- B. E. Conway (1999). Electrochemical Supercapacitors: Scientific Fundamentals and Technological Applications. Berlin: Springer. ISBN 0306457369. Retrieved May 2, 2013.

- Bird, John (2010). Electrical and Electronic Principles and Technology. Routledge. pp. 63–76. ISBN 9780080890562. Retrieved March 17, 2013.

- Spellman, Frank R. (2013). Safe Work Practices for Green Energy Jobs (first ed.). DEStech Publications. p. 323. ISBN 978-1-60595-075-4. Retrieved 29 December 2014

- "Snapshot of Global PV 1992-2014" (PDF). iea-pvps.org. International Energy Agency — Photovoltaic Power Systems Programme. 30 March 2015. Archived from the original on 30 March 2015.

- Hanisch, Christian. "Recycling of Lithium-Ion Batteries" (PDF). Presentation on Recycling of Lithium-Ion Batteries. Lion Engineering GmbH. Retrieved 22 July 2015.

- Delacey, Lynda (October 29, 2015). "Enphase plug-and-play solar energy storage system to begin pilot program". www.gizmag.com. Retrieved December 20, 2015.

- REN21. "Renewables 2014: Global Status Report" (PDF). Archived from the original on 4 September 2014. Retrieved 20 January 2015.

- "Global Market Outlook for Photovoltaics 2014-2018" (PDF). epia.org. EPIA – European Photovoltaic Industry Association. Archived from the original on 12 June 2014. Retrieved 12 June 2014.

- iea.org (2014). "Technology Roadmap: Solar Photovoltaic Energy" (PDF). IEA. Archived from the original on 7 October 2014. Retrieved 7 October 2014.

- Bullis, Kevin (2014-02-18). "How to Make a Cheap Battery for Storing Solar Power | MIT Technology Review". Technologyreview.com. Retrieved 2014-04-27.

- "Notice of the development of mass production technology of Secondary battery "battenice" based on quantum technology" (PDF). MICRONICS JAPAN. 2013-11-19. Retrieved 2014-01-18.

6

An Overview of Nutrient Cycle

This chapter explores the topic of nutrient recycling which is a natural method of exchange of organic and inorganic matter back into production of living matter. Important in this is the phenomenon of nutrition transfer among different organisms. The content provides a comprehensive outline of the topics of complete and closed loop, ecosystem engineers and ecological recycling.

Nutrient Cycle

A nutrient cycle (or ecological recycling) is the movement and exchange of organiand inorganimatter back into the production of living matter. The process is regulated by food web pathways that decompose matter into mineral nutrients. Nutrient cycles occur within ecosystems. Ecosystems are interconnected systems where matter and energy flows and is exchanged as organisms feed, digest, and migrate about. Minerals and nutrients accumulate in varied densities and uneven configurations across the planet. Ecosystems recycle locally, converting mineral nutrients into the production of biomass, and on a larger scale they participate in a global system of inputs and outputs where matter is exchanged and transported through a larger system of biogeochemical cycles.

Composting within agricultural systems capitalizes upon the natural services of nutrient recycling in ecosystems. Bacteria, fungi, insects, earthworms, bugs, and other creatures dig and digest the compost into fertile soil. The minerals and nutrients in the soil is recycled back into the production of crops.

Particulate matter is recycled by biodiversity inhabiting the detritus in soils, water columns, and along particle surfaces (including 'aeolian dust'). Ecologists may refer to ecological recycling, organirecycling, biocycling, cycling, biogeochemical recycling, natural recycling, or just recycling in reference to the work of nature. Whereas the global biogeochemical cycles describe the natural movement and exchange of every kind of particulate matter through the living and non-living components of the Earth, nutrient cycling refers to the biodiversity within community food web systems that loop organinutrients or water supplies back into production. The difference is a matter of scale and compartmentalization with nutrient cycles feeding into global biogeo-chemical cycles. Solar energy flows through ecosystems along unidirectional and non-cyclipathways, whereas the movement of mineral nutrients is cyclic. Mineral cycles include carbon cycle, sulfur cycle, nitrogen cycle, water cycle, phosphorus cycle, oxygen cycle, among others that continually recycle along with other mineral nutrients into productive ecological nutrition. Global biogeochemical cycles are the sum product of localized ecological recycling regulated by the action of food webs moving particulate matter from one living generation onto the next. Earths ecosystems have recycled mineral nutrients sustainably for billions of years.

Outline

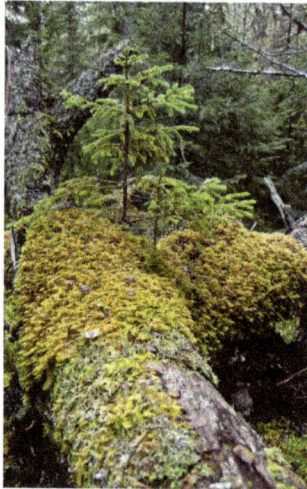

Fallen logs are critical components of the nutrient cycle in terrestrial forests. Nurse logs form habitats for other creatures that decompose the materials and recycle the nutrients back into production.

The nutrient cycle is nature's recycling system. All forms of recycling have feedback loops that use energy in the process of putting material resources back into use. Recycling in ecology is regulated to a large extent during the process of decomposition. Ecosystems employ biodiversity in the food webs that recycle natural materials, such as mineral nutrients, which includes water. Recycling in natural systems is one of the many ecosystem services that sustain and contribute to the well-being of human societies.

There is much overlap between the terms for biogeochemical cycle and nutrient cycle. Most textbooks integrate the two and seem to treat them as synonymous terms. However, the terms often appear independently. Nutrient cycle is more often used in direct reference to the idea of an intra-system cycle, where an ecosystem functions as a unit. From a practical point it does not make sense to assess a terrestrial ecosystem by considering the full column of air above it as well as the great depths of Earth below it. While an ecosystem often has no clear boundary, as a working model it is practical to consider the functional community where the bulk of matter and energy transfer occurs. Nutrient cycling occurs in ecosystems that participate in the "larger biogeochemical cycles of the earth through a system of inputs and outputs."

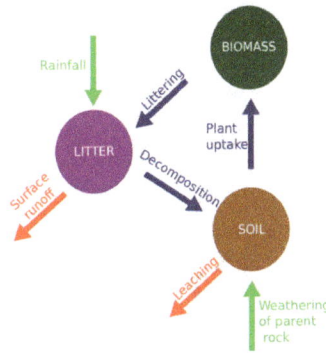

A nutrient cycle of a typical terrestrial ecosystem.

Complete and Closed Loop

All systems recycle. The biosphere is a network of continually recycling materials and information in alternating cycles of convergence and divergence. As materials converge or become more concentrated they gain in quality, increasing their potentials to drive useful work in proportion to their concentrations relative to the environment. As their potentials are used, materials diverge, or become more dispersed in the landscape, only to be concentrated again at another time and place.

Ecosystems are capable of complete recycling. Complete recycling means that 100% of the waste material can be reconstituted indefinitely. This idea was captured by Howard T. Odum when he penned that "it is thoroughly demonstrated by ecological systems and geological systems that all the chemical elements and many organisubstances can be accumulated by living systems from background crustal or oceaniconcentrations without limit as to concentration so long as there is available solar or other source of potential energy" In 1979 Nicholas Georgescu-Roegen proposed a fourth law of entropy stating that complete recycling is impossible. Despite Georgescu-Roegen's extensive intellectual contributions to the science of ecological economics, the fourth law has been rejected in line with observations of ecological recycling. However, some authors state that complete recycling is impossible for technological waste.

Ecosystems execute closed loop recycling where demand for the nutrients that adds to the growth of biomass exceeds supply within that system. There are regional and spatial differences in the rates of growth and exchange of materials, where some ecosystems may be in nutrient debt (sinks) where others will have extra supply (sources). These differences relate to climate, topography, and geological history leaving behind different sources of parent material. In terms of a food web, a cycle or loop is defined as "a directed sequence of one or more links starting from, and ending at, the same species." An example of this is the microbial food web in the ocean, where "bacteria are exploited, and controlled, by protozoa, including heterotrophimicroflagellates which are in turn exploited by ciliates. This grazing activity is accompanied by excretion of substances which are in turn used by the bacteria, so that the system more or less operates in a closed circuit."

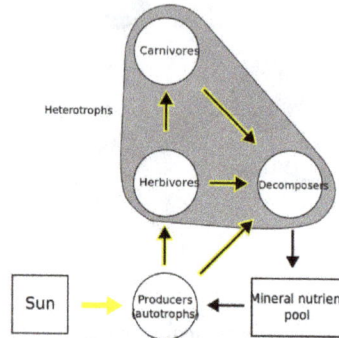

A simplified food web illustrating a three-trophifood chain (*producers-herbivores-carnivores*) linked to decomposers. The movement of mineral nutrients through the food chain, into the mineral nutrient pool, and back into the trophisystem illustrates ecological recycling. The movement of energy, in contrast, is unidirectional and noncyclic.

Ecological Recycling

A large fraction of the elements composing living matter reside at any instant of time in the world's biota. Because the earthly pool of these elements is limited and the rates of exchange among the various components of the biota are extremely fast with respect to geological time, it is quite evident that much of the same material is being incorporated again and again into different biological forms. This observation gives rise to the notion that, on the average, matter (and some amounts of energy) are involved in cycles. An example of ecological recycling occurs in the enzymatidigestion of cellulose. "Cellulose, one of the most abundant organicompounds on Earth, is the major polysaccharide in plants where it is part of the cell walls. Cellulose-degrading enzymes participate in the natural, *ecological recycling* of plant material." Different ecosystems can vary in their recycling rates of litter, which creates a complex feedback on factors such as the competitive dominance of certain plant species. Different rates and patterns of ecological recycling leaves a legacy of environmental effects with implications for the future evolution of ecosystems.

Ecological recycling is common in organifarming, where nutrient management is *fundamentally different* compared to agri-business styles of soil management. Organifarms that employ ecosystem recycling to a greater extent support more species (increased levels of biodiversity) and have a different food web structure. Organiagricultural ecosystems rely on the services of biodiversity for the recycling of nutrients through soils instead of relying on the supplementation of synthetifertilizers. The model for ecological recycling agriculture adheres to the following principals:

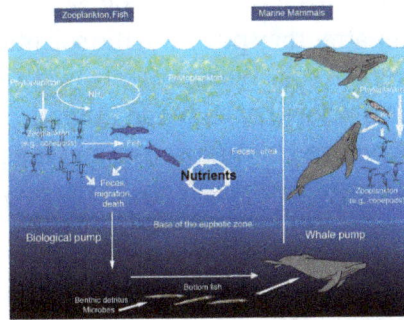

From the largest to the smallest of creatures, nutrients are recycled by their movement, by their wastes, and by their metaboliactivities. This illustration shows an example of the whale pump that cycles nutrients through the layers of the oceaniwater column. Whales can migrate to great depths to feed on bottom fish (such as sand lance *Ammodytes* spp.) and surface to feed on krill and plankton at shallower levels. The whale pump enhances growth and productivity in other parts of the ecosystem.

- Protection of biodiversity.

- Use of renewable energy.

- Recycling of plant nutrients.

Ecosystem Engineers

Fig. 4.

A casting from the Nilgiri Mountains in South India ; of natural size, engraved from a photograph.

An illustration of an earthworm casting taken from Charles Darwin's publication on the movement of organimatter in soils through the ecological activities of worms.

The persistent legacy of environmental feedback that is left behind by or as an extension of the ecological actions of organisms is known as niche construction or ecosystem engineering. Many species leave an effect even after their death, such as coral skeletons or the extensive habitat modifications to a wetland by a beaver, whose components are recycled and re-used by descendants and other species living under a different selective regime through the feedback and agency of these legacy effects. Ecosystem engineers can influence nutrient cycling efficiency rates through their actions.

Earthworms, for example, passively and mechanically alter the nature of soil environments. Bodies of dead worms passively contribute mineral nutrients to the soil. The worms also mechanically modify the physical structure of the soil as they crawl about (bioturbation), digest on the moulds of organimatter they pull from the soil litter. These activities transport nutrients into the mineral layers of soil. Worms discard wastes that create worm castings containing undigested materials where bacteria and other decomposers gain access to the nutrients. The earthworm is employed in this process and the production of the ecosystem depends on their capability to create feedback loops in the recycling process.

Shellfish are also ecosystem engineers because they: 1) Filter suspended particles from the water column; 2) Remove excess nutrients from coastal bays through denitrification; 3) Serve as natural coastal buffers, absorbing wave energy and reducing erosion from boat wakes, sea level rise and storms; 4) Provide nursery habitat for fish that are valuable to coastal economies.

History

Nutrient cycling has a historical foothold in the writings of Charles Darwin in reference to the decomposition actions of earthworms. Darwin wrote about "the continued Following the Greeks, the idea of a hydrological cycle (water is considered a nutrient) was validated and quantified by Halley in 1687

Variations in Terminology

In 1926 Vernadsky coined the term biogeochemistry as a sub-discipline of geochemistry. However, the term nutrient cycle pre-dates biogeochemistry in a pamphlet on silviculture in 1899: "These demands by no means pass over the fact that at places where sufficient quantities of humus are available and where, in case of continuous decomposition of litter, a stable, nutrient humus is present, considerable quantities of nutrients are also available from the biogeninutrient cycle for the standing timber. In 1898 there is a reference to the nitrogen cycle in relation to nitrogen fixing microorganisms. Other uses and variations on the terminology relating to the process of nutrient cycling appear throughout history:

- The term mineral cycle appears early in a 1935 in reference to the importance of minerals in plant physiology: "...ash is probably either built up into its permanent structure, or deposited in some way as waste in the cells, and so may not be free to re-enter the *mineral cycle*."

- The term nutrient recycling appears in a 1964 paper on the food ecology of the wood stork: "While the periodidrying up and reflooding of the marshes creates special survival problems for organisms in the community, the fluctuating water levels favor rapid *nutrient recycling* and subsequent high rates of primary and secondary production":

- The term natural cycling appears in a 1968 paper on the transportation of leaf litter and its chemical elements for consideration in fisheries management:

"Fluvial transport of tree litter from drainage basins is a factor in *natural cycling* of chemical elements and in degradation of the land."

- The term ecological recycling appears in a 1968 publication on future applications of ecology for the creation of different modules designed for living in extreme environments, such as space or under sea: "For our basirequirement of recycling vital resources, the oceans provide much more frequent *ecological recycling* than the land area. Fish and other organipopulations have higher growth rates, vegetation has less capricious weather problems for sea harvesting"

- The term bio-recycling appears in a 1976 paper on the recycling of organicarbon in oceans: "Following the actualistiassumption, then, that biological activity is responsible for the source of dissolved organimaterial in the oceans, but is not important for its activities after death of the organisms and subsequent chemical changes which prevent its *bio-recycling*, we can see no major difference in the behavior of dissolved organimatter between the prebiotiand post-biotioceans."

Water is also a nutrient. In this context, some authors also refer to precipitation recycling, which "is the contribution of evaporation within a region to precipitation in that same region." These variations on the theme of nutrient cycling continue to be used and all refer to processes that are part of the global biogeochemical cycles. However, authors tend to refer to natural, organic, ecological, or bio-recycling in reference to the work of nature, such as it is used in organifarming or ecological agricultural systems.

Recycling in Novel Ecosystems

An endless stream of technological waste accumulates in different spatial configurations across the planet and turns into a predator in our soils, our streams, and our oceans. This idea was similarly expressed in 1954 by ecologist Paul Sears: "We do not know whether to cherish the forest as a source of essential raw materials and other benefits or to remove it for the space it occupies. We expect a river to serve as both vein and artery carrying away waste but bringing usable material in the same channel. Nature long ago discarded the nonsense of carrying poisonous wastes and nutrients in the same vessels." Ecologists use population ecology to model contaminants as competitors or predators. Rachel Carson was an ecological pioneer in this area as her book *Silent Spring* inspired research into biomagification and brought to the worlds attention the unseen pollutants moving into the food chains of the planet.

In contrast to the planets natural ecosystems, technology (or technoecosystems) is not reducing its impact on planetary resources. Only 7% of total plastiwaste (adding up to millions upon millions of tons) is being recycled by industrial systems; the 93% that never makes it into the industrial recycling stream is presumably *absorbed* by natural recycling systems In contrast and over extensive lengths of time (billions of years) ecosystems have maintained a consistent balance with production roughly equaling respiratory consumption rates. The balanced recycling efficiency of nature means that production of decaying waste material has exceeded rates of recyclable consumption into food chains equal to the global stocks of fossilized fuels that escaped the chain of decomposition.

Pesticides soon spread through everything in the ecosphere-both human technosphere and nonhuman biosphere-returning from the 'out there' of natural environments back into plant, animal, and human bodies situated at the 'in here' of artificial environments with unintended, unanticipated, and unwanted effects. By using zoological, toxicological, epidemiological, and ecological insights, Carson generated a new sense of how 'the environment' might be seen.

Microplastics and nanosilver materials flowing and cycling through ecosystems from pollution and discarded technology are among a growing list of emerging ecological concerns. For example, unique assemblages of marine microbes have been found to digest plasti-accumulating in the worlds oceans. Discarded technology is absorbed into soils and creates a new class of soils called technosols. Human wastes in the Anthropocene are creating new systems of ecological recycling, novel ecosystems that have to contend with the mercury cycle and other synthetimaterials that are streaming into the biodegradation chain. Microorganisms have a significant role in the removal of synthetiorganicompounds from the environment empowered by recycling mechanisms that have complex biodegradation pathways. The effect of synthetimaterials, such as nanoparticles and microplastics, on ecological recycling systems is listed as one of the major concerns for ecosystem in this century.

Technological Recycling

Recycling in human industrial systems (or technoecosystems) differs from ecological recycling in scale, complexity, and organization. Industrial recycling systems do not focus on the employment of ecological food webs to recycle waste back into different kinds of marketable goods, but primarily employ people and technodiversity instead. Some researchers have questioned the premise behind these and other kinds of technological solutions under the banner of 'eco-efficiency' are limited in their capability, harmful to ecological processes, and dangerous in their hyped capabilities. Many technoecosystems are competitive and parasititoward natural ecosystems. Food web or biologically based "recycling includes metabolirecycling (nutrient recovery, storage, etc.) and ecosystem recycling (leaching and *in situ* organimatter mineralization, either in the water column, in the sediment surface, or within the sediment."

References

- Lavelle, P.; Dugdale, R.; Scholes, R.; Berhe, A. A.; Carpenter, E.; Codispoti, L.; et al. (2005). "12. Nutrient cycling". Millennium Ecosystem Assessment: Objectives, Focus, and Approach (PDF). Island Press. ISBN 1-55963-228-3.

- Levin, Simon A; Carpenter, Stephen R; Godfray, Charles J; Kinzig, Ann P; Loreau, Michel; Losos, Jonathan B; Walker, Brian; Wilcove, David S (27 July 2009). The Princeton Guide to Ecology. Princeton University Press. p. 330. ISBN 0-691-12839-1.

- Roughgarden, J.; May, R. M.; Levin, S. A. (eds.). "13. Food webs and community structure". Perspectives in ecological theory. Princeton University Press. pp. 181–202. ISBN 0-691-08508-0.

- Mäder, P. "Sustainability of organic and integrated farming (DOK trial)". In Rämert, B.; Salomonsson, L.; Mäder, P. Ecosystem services as a tool for production improvement in organic farming – the role and impact of biodiversity (PDF). Uppsala: Centre for Sustainable Agriculture, Swedish University of Agricultural Sciences. pp. 34–35. ISBN 91-576-6881-7.

Permissions

Index

A
Abandonware, 27, 82-86, 88-91

B
Battery Recycling, 155-157, 159-161
Bio Energy, 201
Biofuel Development, 209
Buy-back Centers, 8-9

C
Chemical Recycling, 14
Clean Mrf, 144
Clearing, 41, 46, 52-55
Closed Loop, 221, 223
Common Land, 27, 118, 120-132
Community Gardening, 27, 101-106
Community Greens, 27, 94-96, 98
Computer Recycling, 13, 26-28, 30, 34, 40-41, 60, 133
Consumer Recycling, 32, 68
Copyright Expiration, 91
Corporate Recycling, 33
Curbside Collection, 8, 20, 134

D
Damage From Cell Reversal, 165
Data Breach, 40, 42-43
Data Erasure, 27, 41-45, 57
Data Remanence, 27, 33, 44, 51-53, 55-56, 58
Data Shredder, 27, 41
Degaussing, 27, 41-43, 47-52, 54-55, 57
Depth of Discharge, 166
Distributed Recycling, 9
Drop-off Centers, 8-9

E
E-waste Management, 66
E-waste Recycling, 13, 27, 30, 63-65
Ecological Recycling, 221-224, 227-228
Electrochemical, 161-162, 164-165, 171, 175, 180-181, 220
Electronic Waste, 4, 27-31, 34-39, 41, 43, 58-63, 65, 68-70
Electronic Waste Substances, 70
Energy Storage, 155, 172-180, 182, 184-191, 203, 208, 218-220

F
File Deletion, 27, 42, 52, 74
Full Disk Overwriting, 44

G
Geothermal Development, 210
Geothermal Energy, 200, 215
Green Alliance, 27, 99-100
Greening, 27, 93-96, 98-99, 101

H
Home Computer Remake, 27, 81
Home Energy Storage, 186
Hydroelectricity, 175-177, 179, 184, 188, 191, 198, 205
Hydropower, 193, 198, 214

I
Interseasonal Thermal Storage, 186

K
Kerbside Collection, 134-139

L
Lead-acid Batteries, 72, 156, 166, 168, 171, 182
Life-cycle Thinking, 151-152
Lithium Ion Batteries, 157, 190

M
Magnetic Data Storage Media, 51
Materials Recovery Facility, 6, 134, 136, 141, 143, 145
Mechanical Storage, 176
Media Destruction, 52, 55
Minimalism (computing), 27, 81
Mixed-waste Processing Facility, 144

N
Nutrient Cycle, 221-223, 225-227, 229

O

Overwriting, 42, 44-46, 52-54, 56-57, 75, 133

P

Park, 27, 94-97, 103, 107, 111-118, 124, 128, 191

Photovoltaic Power Stations, 208

Physical Recycling, 14

Plastic Recycling, 14-15

Purging, 46, 52-53, 55

Q

Quality of Recyclate, 6-7

R

Rechargeable Battery, 160-162, 166, 169, 172-175, 180-181, 186, 190

Recharging Time, 167

Recyclates, 5-10, 17-18

Recycling Codes, 15-16, 147, 149, 151, 153

Recycling Consumer Waste, 7

Recycling in Novel Ecosystems, 227

Recycling Industrial Waste, 11

Renewable Energy, 12, 155, 163, 174, 187-188, 191-194, 196-197, 203, 205, 209, 211-216, 219, 225

Resin Identification Code, 147-148, 154

Retrocomputing, 27, 75-76, 81

Rinsing, 11

S

Scrapping, 32

Secure Recycling, 40

Silver Oxide Batteries, 157

Single-stream Recycling, 10, 25, 134, 141

Solar Energy, 183-184, 187, 190, 195-197, 199, 220, 222

Solar Thermal, 187, 199, 207-208, 217

Sorting, 7-11, 22-23, 140, 142, 144-145, 153, 157

Source Reduction, 134, 145-146, 153

T

Table of Resin Codes, 149

Technological Recycling, 228

Thermal Storage, 179, 186, 188-189

Trade in Recyclates, 18

V

Village Green, 27, 107-111

W

Waste Hierarchy, 1, 147, 150-154

Waste Plastic Pyrolysis to Fuel Oil, 15

Wet Mrf, 145

Wind Power, 174, 180, 192, 194, 197, 203, 205-207, 214, 216, 218-219

Wind Power Development, 206

www.ingramcontent.com/pod-product-compliance
Lightning Source LLC
Chambersburg PA
CBHW061947190326
41458CB00009B/2805